MATH AMUSEMENTS
IN
DEVELOPING SKILLS

by

Alice A. Clack and Carol H. Leitch

Midwest Publications Company, Inc.
P.O. Box 448
Pacific Grove, CA 93950

Preface

    The materials provided in this volume are designed to supplement the elementary school curriculum. They can also be used as review or remedial materials for middle and high school students.

    The student solves various types of puzzles including cross number puzzles, dot-to-dots, decode, and design puzzles, which are interesting and enjoyable, making necessary practice in basic skills a more appealing experience.

    A section of holiday puzzles is included to aid in making those pre-vacation days of the school year more enjoyable for teacher and student.

Table of Contents

Page

READING AND WRITING WHOLE NUMBERS

Connect these numbers in order.

1. seventeen
2. twenty-three
3. fifty-six
4. nine
5. fourteen
6. eighty-three
7. five
8. thirty-nine
9. one hundred
10. ninety-eight
11. seventy-seven
12. five hundred eighty-one
13. one hundred fifteen
14. twenty-seven
15. two hundred thirty
16. eighty-five
17. eleven
18. forty-one
19. sixty-nine
20. eight
21. ninety-four
22. six hundred twenty-two
23. thirteen
24. eight hundred
25. forty-two
26. twelve
27. seven
28. four hundred nine
29. fifty-eight
30. seven hundred ninety
31. twenty-six
32. seventy-three
33. nine hundred six
34. thirty-four
35. five hundred eighty-one
36. three hundred

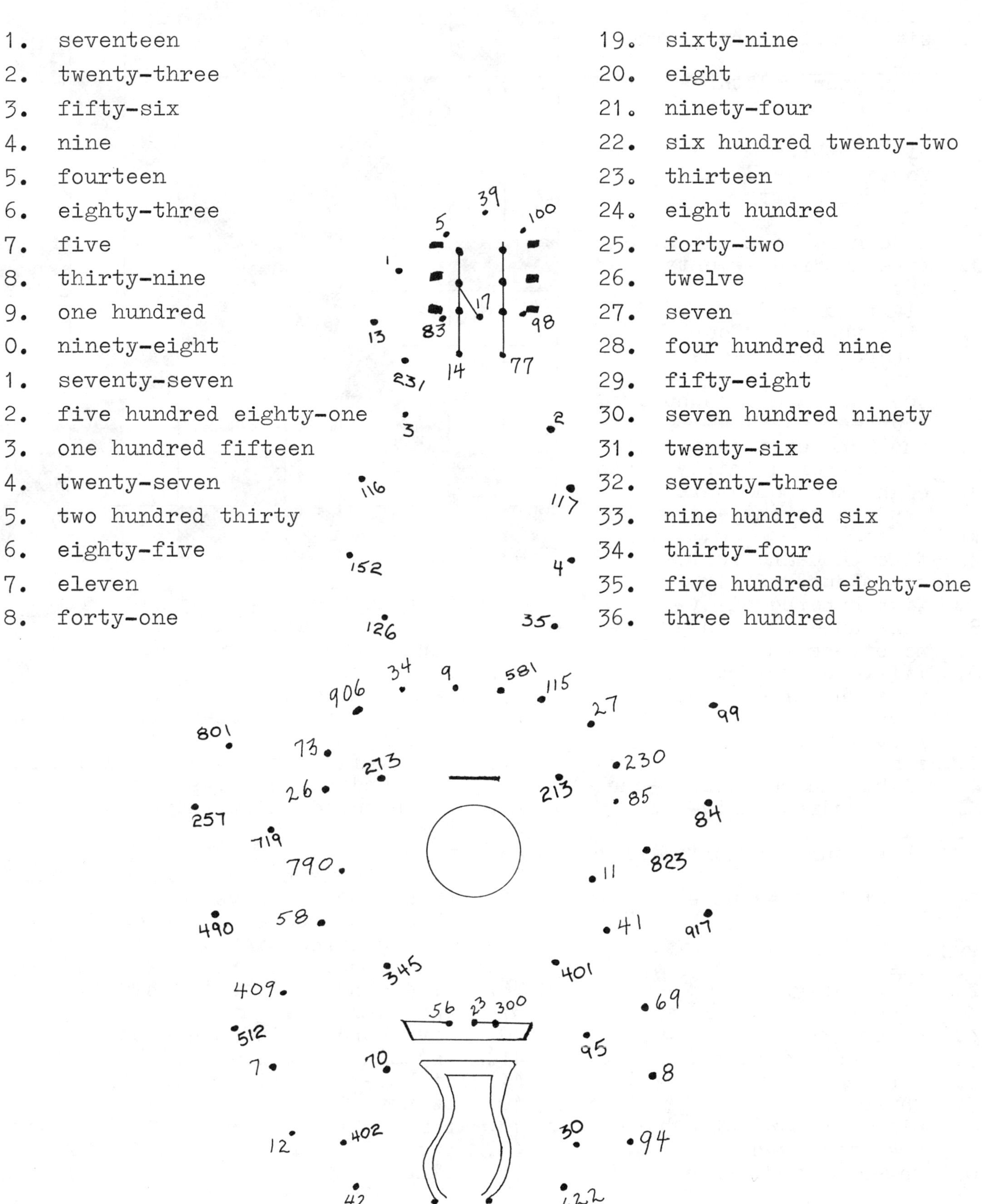

Across

1. one hundred thirty-five
4. seven
5. six hundred forty-three
8. nine
9. eight hundred three
11. four hundred forty
12. eight
13. ninety-three
15. forty thousand three hundred eighty
17. twenty
18. nine hundred sixty-nine
20. five hundred seventy-three
21. eight hundred
22. nine thousand four hundred thirty-one
24. two
25. two thousand five hundred fourteen
26. one hundred twenty-four
28. nine hundred thirty-seven
29. eight thousand three hundred seventy-six
31. three
32. five thousand forty-six
35. seven hundred six
36. five hundred sixty-two
38. nine hundred eighty
39. ninety-seven
40. fifty-seven thousand three hundred eighty-seven
42. forty-three
43. six
44. eight hundred twenty-one
45. six hundred sixty-two
47. six
48. three hundred forty-three
49. two
50. eight hundred nine

Down

1. one
2. thirty-eight
3. five hundred four
4. seven
5. six hundred forty
6. forty
7. three
8. nine thousand nine hundred ninety-nine
10. three hundred five
11. four hundred eighty-three
12. eight thousand four
14. three hundred sixty-four
16. three hundred seventy-two
17. two hundred one
19. ninety-three thousand one hundred seventy-six
21. eighty-five thousand seven hundred nin
23. one hundred twenty-six
25. two hundred thirty-five
27. four
28. nine
29. eight thousand seven hundred ninety-si
30. three hundred seven
31. three hundred sixty-three
33. four hundred eighty-four
34. six thousand thirty-six
36. five hundred seventy-one
37. two hundred eighty-six
40. five hundred twenty-three
41. seven hundred sixty-eight
44. eighty-four
46. twenty
48. three
49. two
51. nine

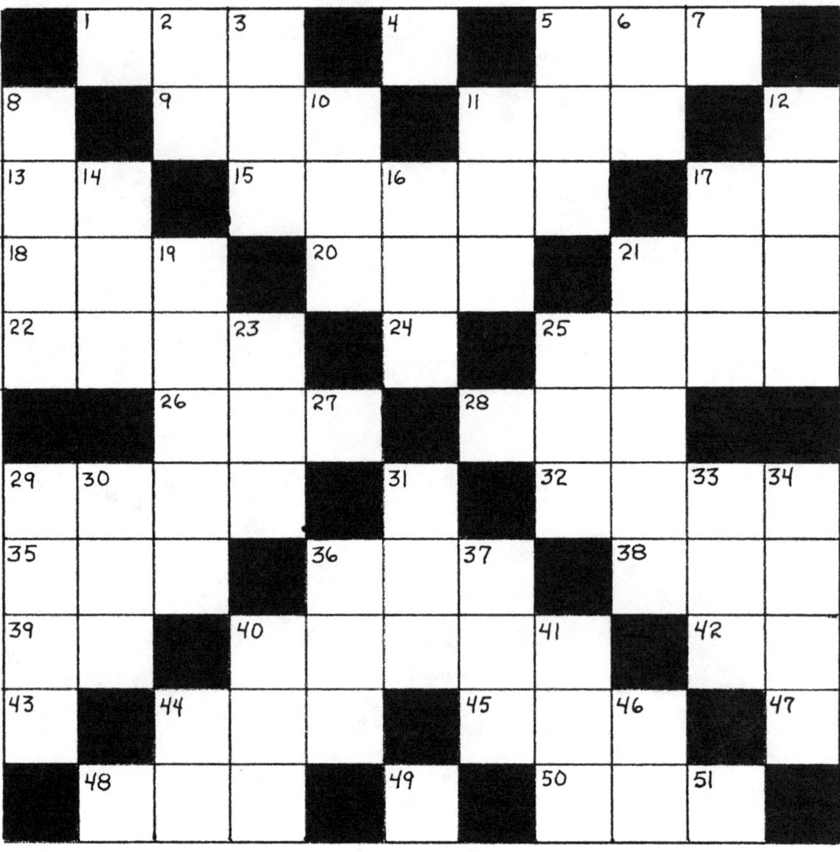

ADDITION OF WHOLE NUMBERS

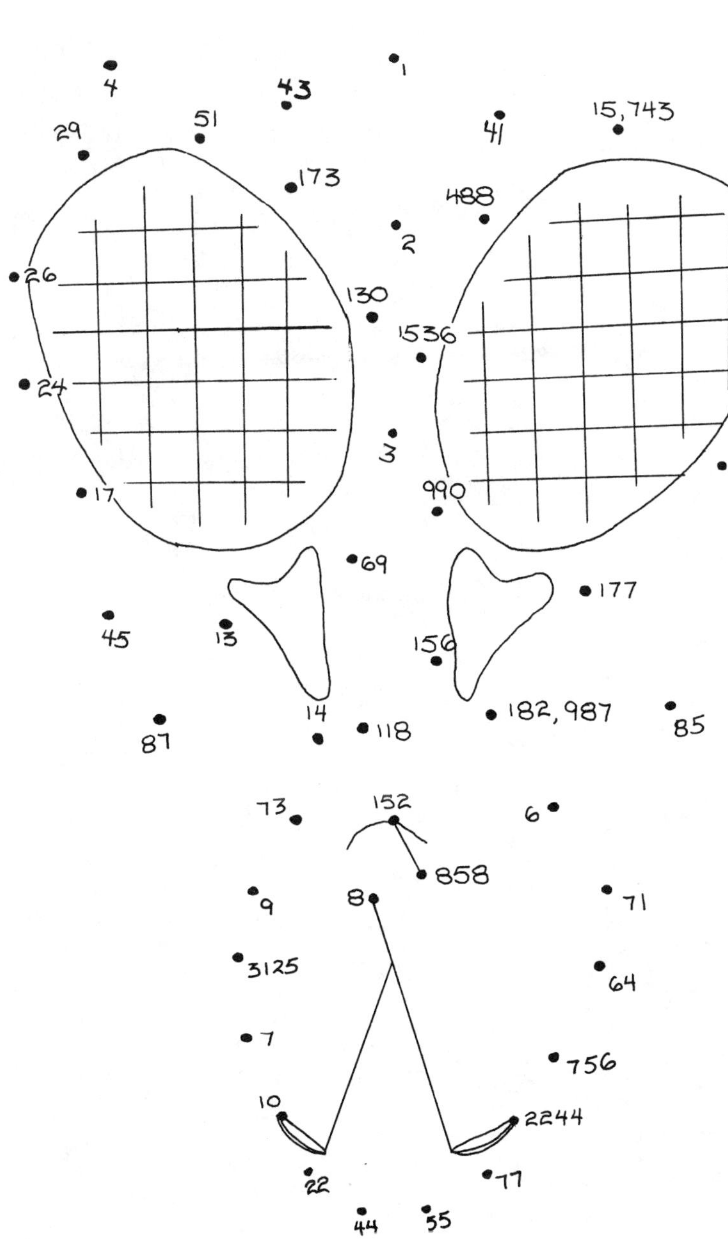

Connect the answers in order.

1. 5 + 2 + 3 =   ______
2. 2 + 4 + 2 =   ______
3. 7 + 6 + 1 =   ______
4. 4 + 6 + 3 =   ______
5. 8 + 7 + 2 =   ______
6. 9 + 9 + 6 =   ______
7. 6 + 8 + 4 + 8 =   ______
8. 9 + 2 + 7 + 3 + 8 = ______
9. 16 + 35 =   ______
10. 97 + 76 =   ______
11. 50 + 80 =   ______
12. 11 + 23 + 35 =   ______
13. 67 + 20 + 31 =   ______
14. 67 + 27 + 58 =   ______
15. 32 + 45 + 67 + 12 = ______
16. 426 + 564 =   ______
17. 609 + 927 =   ______
18. 437 + 46 + 5 =   ______
19. 7648 + 8095 =   ______
20. 56 + 4 + 27 + 7 =   ______
21. 537 + 80 + 9 + 347 = ______
22. 52 + 349 + 76 =   ______
23. 25 + 38 + 90 + 24 = ______
24. 97,508 + 85,479 =   ______
25. 785 + 64 + 9 =   ______
26. 865 + 437 + 942 =   ______

· 135

81 ·

7·

9·

· 26

87 ·

· 59

6 ·

84·

·28

· 14

· 131

Connect answer "a" with answer "b" for each problem.

	(a)		(b)
1.	$2 + 4 + 3 + 4 + 1 =$ ______	$4 + 8 + 4 + 9 + 3 =$ ______
2.	$5 + 0 + 3 + 1 =$ ______	$17 + 36 + 22 + 9 =$ ______
3.	$18 + 27 + 36 =$ ______	$2 + 3 + 1 + 1 =$ ______
4.	$2 + 0 + 3 + 1 =$ ______	$58 + 17 + 9 + 3 =$ ______
5.	$9 + 7 + 5 + 8 + 30 =$ ______	$2 + 6 + 7 + 6 + 5 =$ ______
6.	$32 + 17 + 35 + 3 =$ ______	$15 + 13 + 9 + 22 =$ ______
7.	$28 + 34 + 56 + 17 =$ ______	$2 + 0 + 3 + 4 =$ ______
8.	$56 + 9 + 30 + 19 + 17 =$ ______	$6 + 7 + 12 + 2 + 1 =$ ______
9.	$1 + 0 + 4 + 0 + 2 =$ ______	$5 + 3 + 9 + 8 + 3 =$ ______
10.	$28 + 35 + 42 + 26 =$ ______	$12 + 26 + 34 + 12 =$ ______
11.	$5 + 0 + 1 + 1 =$ ______	$53 + 52 + 30 =$ ______
12.	$22 + 24 + 31 + 7 =$ ______	$2 + 1 + 2 + 1 =$ ______
13.	$5 + 1 + 0 + 3 =$ ______	$9 + 8 + 7 + 2 =$ ______
14.	$52 + 17 + 12 =$ ______	$18 + 21 + 4 + 9 + 35 =$ ______
15.	$5 + 3 + 2 + 4 =$ ______	$12 + 27 + 18 + 2 =$ ______

ADDITION OF WHOLE NUMBERS

Across

1. 23 + 34 + 65
4. 2 + 5 + 1
5. 523 + 46 + 173
8. 3 + 3 + 2 + 1
9. 273 + 519
11. 56 + 143 + 78
12. 8 + 0 + 1
13. 23 + 9 + 63
15. 9999 + 3840
17. 57 + 9 + 8 + 12
18. 257 + 29 + 386
20. 582 + 8 + 76
21. 79 + 56 + 83 + 92
22. 583 + 140 + 196
24. 1 + 5 + 0 + 1 + 2
25. 368 + 144 + 257
26. 2 + 0 + 3 + 4 + 0
27. 58 + 73 + 29 + 62
29. 138 + 406 + 329
30. 2 + 3 + 2 + 1
31. 189 + 276 + 354
32. 4 + 2 + 3 + 0
33. 28 + 73 + 42 + 93
35. 46 + 238 + 97
36. 9 + 87 + 695
38. 483 + 159 + 96
40. 7 + 8 + 9 + 6 + 5
41. 2934 + 85692 + 562
43. 56 + 8 + 17 + 9
44. 1 + 0 + 3 + 5
45. 57 + 93 + 82
46. 203 + 576 + 29
48. 5 + 1 + 0 + 2
49. 83 + 92 + 147
50. 2 + 5 + 1
51. 156 + 98 + 342

Down

1. 1 + 0
2. 9 + 9 + 9
3. 56 + 73 + 162
4. 2 + 3 + 3
5. 283 + 362 + 134
6. 15 + 3 + 9 + 18 + 2
7. 1 + 0 + 1 + 0
8. 238 + 374 + 384
10. 87 + 65 + 49 + 35
11. 34 + 50 + 60 + 92
12. 530 + 104 + 326
14. 98 + 250 + 89 + 142
16. 800 + 60 + 9
17. 204 + 360 + 255
19. 15,386 + 5825
21. 21,403 + 14,934
23. 286 + 501 + 142
25. 75 + 155 + 304 + 238
26. 1 + 1 + 4 + 2 + 1
28. 1 + 0 + 1
29. 5 + 3
30. 4 + 1 + 2 + 1
31. 442 + 443
32. 301 + 502 + 188
34. 600 + 30 + 9
35. 113 + 113 + 113
36. 562 + 230
37. 94 + 47 + 47
39. 80 + 728
41. 250 + 307 + 275
42. 800 + 5
45. 7 + 6 + 4 + 5
47. 16 + 18 + 32 + 23
49. 1 + 1 + 1
50. 2 + 2 + 2 + 2
52. 1 + 0 + 2 + 0 + 3

$$\overline{10}\ \overline{7}\ \overline{7}\ \overline{8}\ \overline{3}\ \overline{8}\ \overline{4}\ \overline{1} \qquad \overline{11}\ \overline{5}\ \overline{3}\ \overline{9} \qquad \overline{8}\ \overline{3} \qquad \overline{3}\ \overline{4}\ \overline{11}\ \overline{5}\ \overline{3}\ \overline{6}\ \overline{5}\ \overline{2}\ !$$

Solve the problems. Find the number that each letter represents.
Fill in the blank with the letters each number represents.

1. 52 + 76 + 34 + 82 = $\underset{R}{\underline{}}$ _ _

2. 923 + 784 + 962 = _ _ $\underset{H}{\underline{}}$ _

3. 83 + 9 + 284 + 73 = _ _ $\underset{S}{\underline{}}$

4. 576 + 93 + 2845 = _ _ $\underset{N}{\underline{}}$ _

5. 4057 + 23 + 420 = $\underset{O}{\underline{}}$ _ _ _

6. 576 + 329 + 450 = _ _ _ $\underset{E}{\underline{}}$

7. 9284 + 28 + 736 + 9 = $\underset{[A]}{\underline{}}$ _ _ _

8. 73 + 2904 + 8 + 134 + 8670 = _ _ _ $\underset{I}{\underline{}}$ _

9. 92,348 + 52,763 = _ _ _ _ $\underset{[G]}{\underline{}}$

10. 19,736 + 52 + 8346 = _ _ _ $\underset{T}{\underline{}}$ _

11. 928 + 43 + 5 + 2806 = _ $\underset{D}{\underline{}}$ _ _

Across

1. 76 - 43
3. 69 - 21
5. 12 - 9
6. 426 - 105
8. 28 - 22
9. 47 - 29
11. 22 - 15
12. 95 - 67
13. 234 - 192
15. 536 - 522
16. 283 - 190
17. 300 - 297
18. 59 - 35
20. 46 - 37
21. 284 - 177
23. 148 - 139
24. 70 - 53
25. 286 - 251

Down

1. 40 - 37
2. 71 - 38
3. 86 - 45
4. 527 - 519
5. 375 - 344
7. 81 - 54
8. 92 - 24
10. 971 - 128
12. 536 - 294
14. 41 - 39
15. 100 - 99
16. 234 - 135
17. 47 - 17
19. 293 - 244
21. 88 - 71
22. 193 - 120
24. 570 - 569
26. 233 - 228

Connect the answers in order.

1. 76 − 34 =   ______
2. 54 − 36 =   ______
3. 73 − 9 =   ______
4. 81 − 49 =   ______
5. 53 − 11 =   ______
6. 70 − 62 =   ______
7. 438 − 426 =   ______
8. 648 − 639 =   ______
9. 763 − 668 =   ______
10. 450 − 449 =   ______
11. 7436 − 7379 =   ______
12. 400 − 376 =   ______
13. 303 − 165 =   ______
14. 1005 − 937 =   ______
15. 769 − 543 =   ______
16. 9284 − 8362 =   ______
17. 49182 − 48520 =   ______
18. 5706 − 4873 =   ______
19. 5000 − 3642 =   ______
20. 4300 − 3400 =   ______
21. 4060 − 4028 =   ______

SUBTRACTION OF WHOLE NUMBERS

The first man to set foot on the moon:

$$\overline{14}\ \overline{12}\ \overline{4}\ \overline{6}\qquad \overline{11}\ \overline{2}\ \overline{1}\ \overline{8}\ \overline{7}\ \overline{2}\ \overline{5}\ \overline{14}\ \overline{3}$$

A famous Chinese teacher:

$$\overline{13}\ \overline{5}\ \overline{14}\ \overline{9}\ \overline{10}\ \overline{13}\ \overline{4}\ \overline{10}\ \overline{8}$$

Solve the problems. Find the number that each letter represents.
Fill in the blank with the letters each number represents.

1.   75 – 63 = _
        ⌐E⌐

2.   95 – 78 = _ _
        M̅

3.   136 – 129 = _
         T̅

4.   546 – 496 = _ _
         O̅

5.   2836 – 1973 = _ _ _
            L̅

6.   82,467 – 81,579 = _ _ _
                  S̅

7.   2897 – 2654 = _ _ _
               G̅

8.   8034 – 7825 = _ _ _
            R̅

9.   2947 – 2863 = _ _
             I̅

10.   98,283 – 86,297 = _ _ _ _ _
                   ⌐A⌐

11.   29,293 – 14,315 = _ _ _ _ _
                   ⌐N⌐

12.   5340 – 2727 = _ _ _ _
                ⌐C⌐

13.   8042 – 1936 = _ _ _ _
             ⌐U⌐

14.   897 – 558 = _ _ _
              F̅

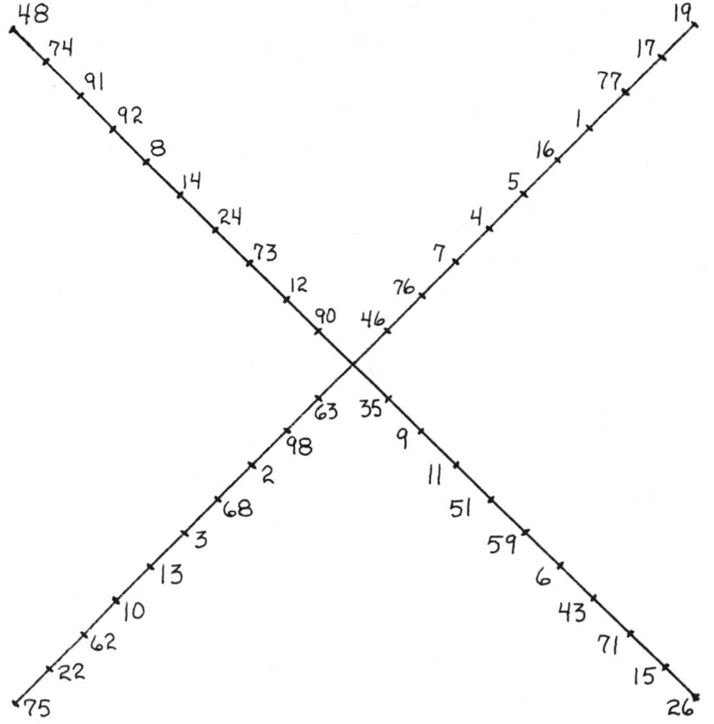

Connect "a" and "b" answers for each of the following problems:

	(a)		(b)	
1.	28 − 17 =	_____	249 − 172 =	_____
2.	86 − 72 =	_____	510 − 497 =	_____
3.	159 − 83 =	_____	42 − 27 =	_____
4.	869 − 777 =	_____	804 − 736 =	_____
5.	4289 − 4273 =	_____	219 − 160 =	_____
6.	530 − 482 =	_____	903 − 840 =	_____
7.	1543 − 1492 =	_____	258 − 257 =	_____
8.	2836 − 2817 =	_____	5340 − 5305 =	_____
9.	629 − 583 =	_____	283 − 257 =	_____
10.	781 − 773 =	_____	435 − 432 =	_____
11.	913 − 842 =	_____	2843 − 2836 =	_____
12.	1035 − 973 =	_____	462 − 389 =	_____
13.	507 − 432 =	_____	2436 − 2346 =	_____
14.	265 − 243 =	_____	899 − 887 =	_____
15.	500 − 457 =	_____	969 − 965 =	_____
16.	4271 − 4269 =	_____	531 − 440 =	_____
17.	2009 − 1999 =	_____	54300 − 54276 =	_____
18.	346 − 340 =	_____	872 − 867 =	_____
19.	4205 − 4196 =	_____	765 − 748 =	_____
20.	3629 − 3555 =	_____	14397 − 14299 =	_____

272

162

108

684

1270

92

756

228

196

96

290

664

Connect "a" and "b" answers for each problem. Then cut out
the triangles formed and put them together to form a parallelogram.

		(a)			(b)
1.	46 x 2 =	_____	34 x 8 =	_____
2.	57 x 4 =	_____	83 x 8 =	_____
3.	36 x 3 =	_____	76 x 9 =	_____
4.	58 x 5 =	_____	32 x 3 =	_____
5.	49 x 4 =	_____	166 x 4 =	_____
6.	27 x 6 =	_____	136 x 2 =	_____
7.	16 x 6 =	_____	254 x 5 =	_____
8.	127 x 10 =	_____	29 x 10 =	_____
9.	4 x 23 =	_____	18 x 9 =	_____
10.	228 x 3 =	_____	252 x 3 =	_____
11.	114 x 2 =	_____	28 x 7 =	_____
12.	84 x 9 =	_____	18 x 6 =	_____

MULTIPLICATION OF WHOLE NUMBERS

A famous Apache Indian, born in Arizona:

$$\overline{10}\ \overline{5}\ \overline{8}\ \overline{7}\ \overline{4}\ \overline{1}\ \overline{9}\ \overline{7}$$

A Sioux Indian Chief, born in South Dakota:

$$\overline{11}\ \overline{1}\ \overline{2}\ \overline{2}\ \overline{1}\ \overline{4}\ \overline{10}\qquad \overline{6}\ \overline{3}\ \overline{12}\ \overline{12}$$

Solve the problems. Find the number that each letter represents.
Fill in the blank with the letters each number represents.

1. 56 x 4 = _ _ _
           N

2. 37 x 5 = _ _ _
           I

3. 87 x 6 = _ _ _
           E

4. 99 x 2 = _ _ _
             R

5. 47 x 7 = _ _ _
           U

6. 313 x 3 = _ _ _
            M

7. 407 x 8 = _ _ _ _
              T

8. 49 x 26 = _ _ _ _
             ⌐L⌐

9. 70 x 16 = _ _ _ _
             ⌐S⌐

10. 8 x 3 x 4 = _ _
                B

11. 5368 x 2 = _ _ _ _ _
               ⌐G⌐

12. 8734 x 9 = _ _ _ _ _
               O

Connect the answers in order.

1.   27 x 8 =   _______
2.   43 x 6 =   _______
3.   89 x 5 =   _______
4.   56 x 9 =   _______
5.   126 x 7 =   _______
6.   248 x 4 =   _______
7.   437 x 3 =   _______
8.   896 x 2 =   _______
9.   564 x 5 =   _______
10. 1234 x 3 =   _______
11. 2571 x 4 =   _______
12. 72 x 3 =   _______
13. 4630 x 9 =   _______
14. 2468 x 0 =   _______
15. 97 x 1 =   _______
16. 58 x 3 =   _______
17. 19 x 5 =   _______
18. 83 x 92 =   _______
19. 95 x 43 =   _______
20. 275 x 301 =   _______
21. 430 x 562 =   _______
22. 911 x 203 =   _______
23. 404 x 376 =   _______
24. 827 x 941 =   _______
25. 119 x 357 =   _______
26. 5 x 4 x 2 x 7 = _______
27. 3 x 6 x 1 x 9 = _______
28. 5 x 8 x 6 x 4 = _______
29. 21 x 8 x 3 =   _______

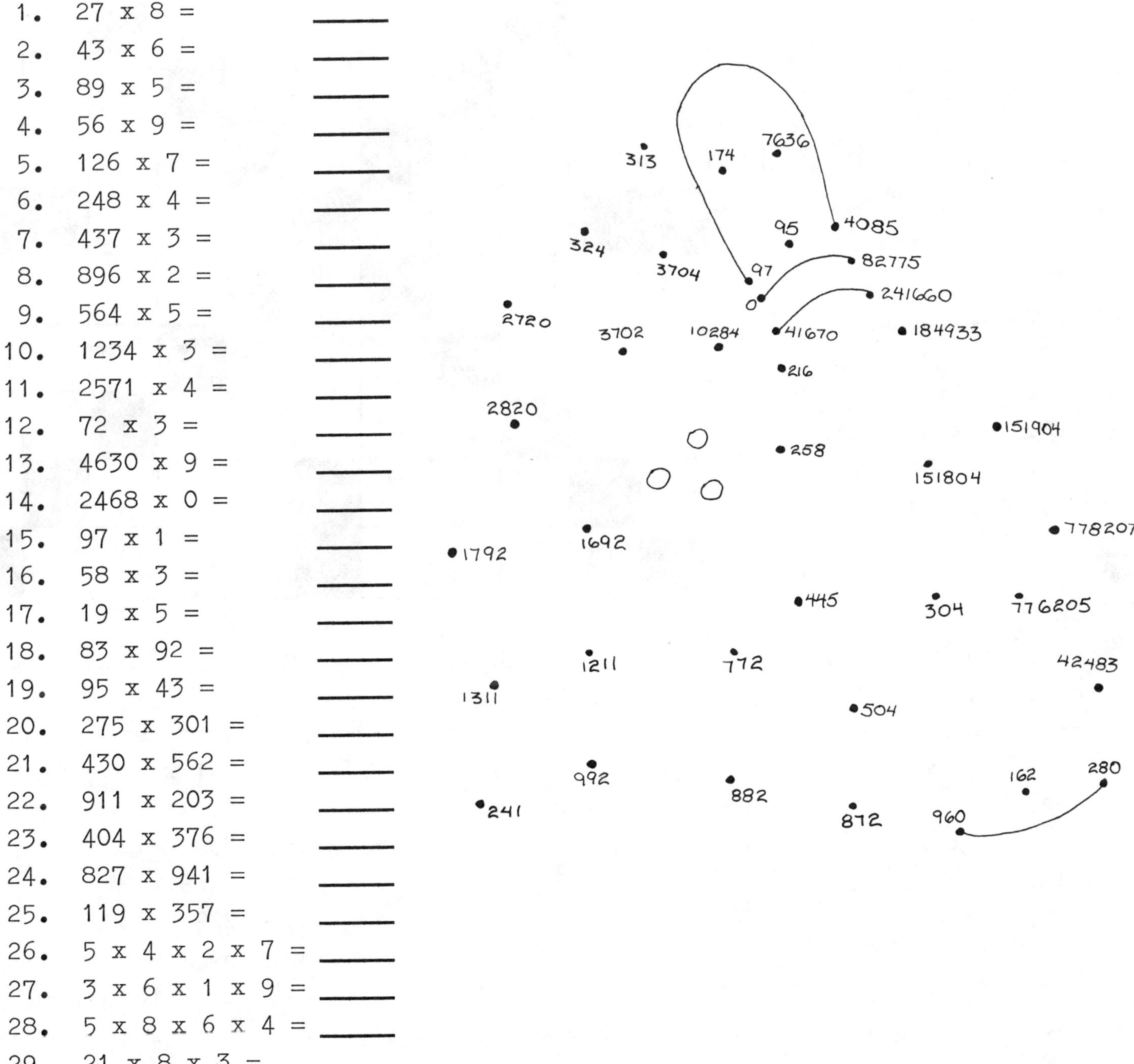

Across

1. 749 x 2
5. 39 x 62
9. 2 x 4
10. 325 x 3
11. 37 x 5
12. 3 x 3
13. 27 x 3
15. 3120 x 8
17. 49 x 2
18. 123 x 7
20. 23 x 34
21. 17 x 26
22. 4137 x 2
24. 2 x 3
25. 56 x 43
26. 8 x 7 x 4 x 2
28. 2 x 3 x 7 x 5
29. 83 x 44
31. 1 x 7
32. 157 x 16
35. 6 x 5 x 3 x 2
36. 33 x 19
38. 142 x 4
39. 14 x 7
40. 343 x 71
42. 4 x 2 x 7
43. 5 x 0
44. 15 x 25
45. 39 x 15
47. 8 x 1
48. 23 x 14 x 9
49. 52 x 36 x 4

Down

1. 1 x 1
2. 7 x 7
3. 486 x 2
4. 407 x 21
5. 1081 x 2
6. 16 x 30
7. 5 x 3
8. 2 x 4
9. 2222 x 4
12. 273 x 36
14. 18 x 9
16. 493 x 2
17. 47 x 20
19. 349 x 50
21. 979 x 45
23. 17 x 26
25. 53 x 4
27. 2 x 2 x 2
28. 2 x 1
29. 319 x 10
30. 43 x 16
31. 241 x 3
33. 15 x 11
34. 239 x 12
36. 3229 x 2
37. 229 x 33
40. 31 x 9
41. 24 x 16
44. 19 x 2
46. 29 x 2
48. 2 x 1
50. 4 x 2

14

$$\overline{6}\ \overline{5}\ \overline{8}\ \overline{5}\ \overline{6}\ \overline{10} \qquad \overline{3}\ \overline{11}\ \overline{6} \qquad \overline{2}\ \overline{9}\ \overline{11}\ \overline{1}\ \overline{4}\ \overline{10}\ \overline{7}\ !$$

Solve the problems. Find the number that each letter represents.
Fill in the blank with the letters each number represents.

1. $320 \div 8 = \underset{\overline{U}}{\quad} \,\underline{\ \ }$

2. $420 \div 6 = \underset{\overline{R}}{\quad} \,\underline{\ \ }$

3. $360 \div 40 = \underset{\overline{O}}{\quad}$

4. $730 \div 73 = \underset{\ulcorner E \urcorner}{\quad}$

5. $75 \div 5 = \underline{\ \ }\ \underset{\overline{I}}{\quad}$

6. $88 \div 8 = \underset{\ulcorner N \urcorner}{\quad}$

7. $305 \div 5 = \underline{\ \ }\ \underset{\overline{Q}}{\quad}$

8. $249 \div 3 = \underline{\ \ }\ \underset{\overline{A}}{\quad}$

9. $256 \div 4 = \underset{\overline{D}}{\quad} \,\underline{\ \ }$

10. $378 \div 9 = \underline{\ \ }\ \underset{\overline{C}}{\quad}$

11. $348 \div 6 = \underline{\ \ }\ \underset{\overline{V}}{\quad}$

DIVISION OF WHOLE NUMBERS

65

32 .

403

.93

17 .

.4

87 .

79 .

10 .   . 95

954

.7

. 36

Connect "a" and "b" answers for each problem. Then cut
out the polygons formed and put them together to form a
rectangle.

(a) (b)

1.   96 ÷ 3 =   _______   360 ÷ 90 =   _______
2.   325 ÷ 5 =   _______   2862 ÷ 3 =   _______
3.   174 ÷ 2 =   _______   430 ÷ 43 =   _______
4.   651 ÷ 7 =   _______   3627 ÷ 9 =   _______
5.   144 ÷ 4 =   _______   490 ÷ 70 =   _______
6.   711 ÷ 9 =   _______   7632 ÷ 8 =   _______
7.   790 ÷ 79 =   _______   180 ÷ 5 =   _______
8.   570 ÷ 6 =   _______   837 ÷ 9 =   _______
9.   560 ÷ 80 =   _______   522 ÷ 6 =   _______
10. 2821 ÷ 7 =   _______   380 ÷ 4 =   _______
11. 352 ÷ 11 =   _______   255 ÷ 15 =   _______
12. 650 ÷ 10 =   _______   1580 ÷ 20 =   _______
13. 510 ÷ 30 =   _______   160 ÷ 40 =   _______

16

Across

1. 115 ÷ 5
3. 108 ÷ 9
5. 468 ÷ 2
8. 81 ÷ 9
9. 49 ÷ 7
10. 332 ÷ 4
11. 3090 ÷ 6
12. 168 ÷ 3
13. 696 ÷ 8
14. 572 ÷ 2
15. 2415 ÷ 5
16. 730 ÷ 73
17. 8757 ÷ 9
18. 5852 ÷ 11
19. 128 ÷ 64
20. 3120 ÷ 20
21. 14000 ÷ 50
22. 265 ÷ 53
23. 654 ÷ 3
24. 15480 ÷ 60
25. 204 ÷ 12
26. 4752 ÷ 11
27. 1183 ÷ 7
28. 5700 ÷ 20
29. 875 ÷ 25
30. 1920 ÷ 16
31. 9500 ÷ 19
32. 141 ÷ 47
33. 3080 ÷ 5
34. 6300 ÷ 15
35. 288 ÷ 72
36. 645 ÷ 3
37. 4662 ÷ 6
38. 625 ÷ 25
39. 508 ÷ 4
40. 20000 ÷ 40
41. 2936 ÷ 8

Down

1. 135 ÷ 5
2. 39 ÷ 13
3. 52 ÷ 4
4. 38 ÷ 19
5. 1080 ÷ 5
6. 245 ÷ 7
7. 68 ÷ 17
8. 5778 ÷ 6
10. 870 ÷ 10
11. 4664 ÷ 8
12. 1746 ÷ 3
13. 1200 ÷ 15
14. 3036 ÷ 11
15. 3010 ÷ 7
16. 144 ÷ 12
17. 4790 ÷ 5
18. 11760 ÷ 20
20. 1008 ÷ 9
21. 518 ÷ 2
22. 17250 ÷ 30
23. 1410 ÷ 6
24. 2080 ÷ 8
25. 2160 ÷ 12
26. 17320 ÷ 40
27. 3150 ÷ 25
28. 1000 ÷ 5
30. 230 ÷ 2
31. 5270 ÷ 10
33. 1851 ÷ 3
34. 9400 ÷ 20
35. 3199 ÷ 7
36. 242 ÷ 11
37. 700 ÷ 10
38. 338 ÷ 13
39. 426 ÷ 426
40. 215 ÷ 43
41. 57 ÷ 19

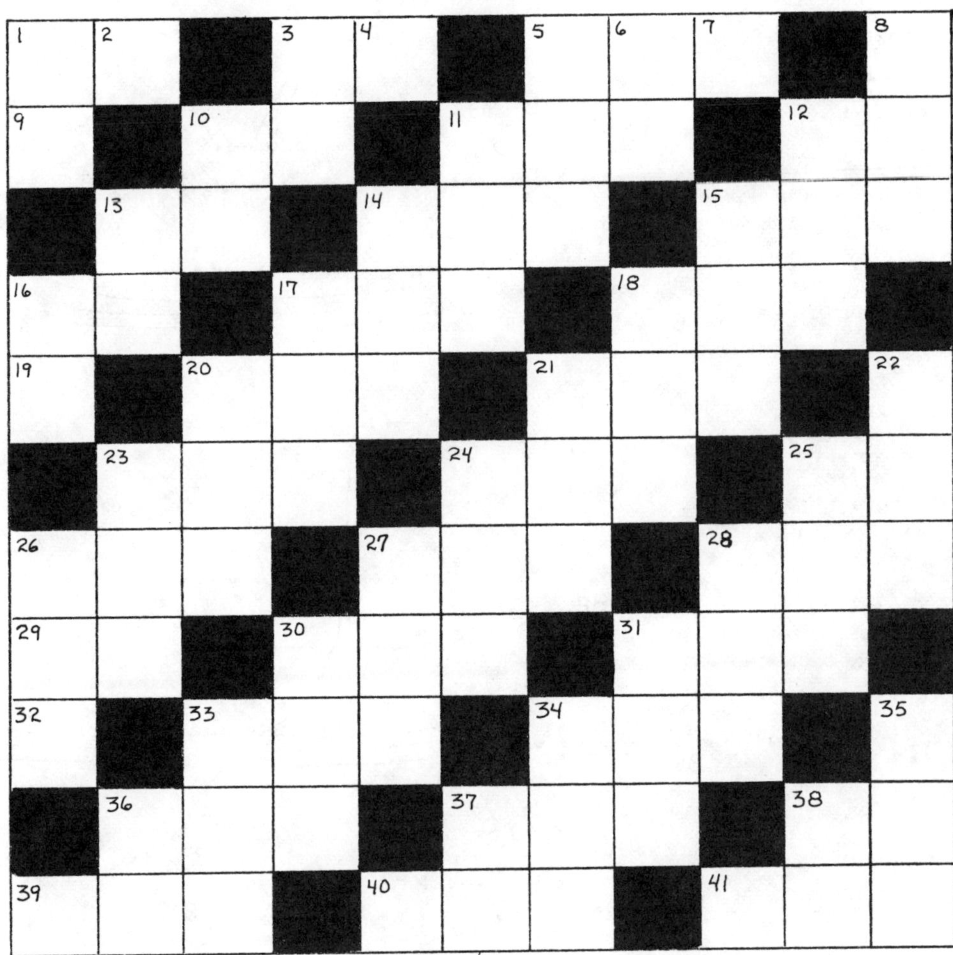

DIVISION OF WHOLE NUMBERS

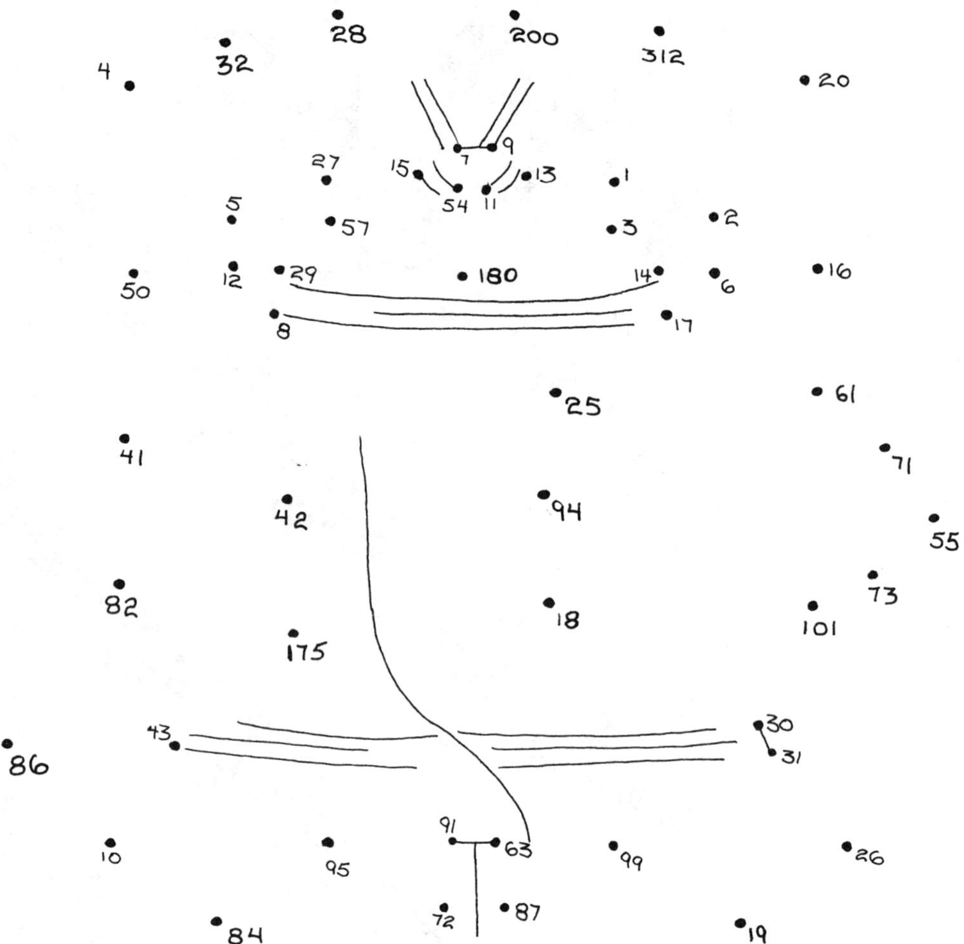

Connect the answers in order.

1. $217 \div 7 =$	18. $175 \div 35 =$
2. $104 \div 4 =$	19. $480 \div 40 =$
3. $99 \div 1 =$	20. $1250 \div 25 =$
4. $126 \div 2 =$	21. $304 \div 76 =$
5. $435 \div 5 =$	22. $1140 \div 57 =$
6. $720 \div 10 =$	23. $224 \div 14 =$
7. $546 \div 6 =$	24. $114 \div 19 =$
8. $285 \div 3 =$	25. $196 \div 98 =$
9. $670 \div 67 =$	26. $573 \div 573 =$
10. $860 \div 20 =$	27. $1040 \div 80 =$
11. $136 \div 17 =$	28. $225 \div 25 =$
12. $261 \div 9 =$	29. $495 \div 45 =$
13. $1881 \div 33 =$	30. $297 \div 99 =$
14. $378 \div 7 =$	31. $1106 \div 79 =$
15. $315 \div 45 =$	32. $680 \div 40 =$
16. $450 \div 30 =$	33. $1170 \div 39 =$
17. $324 \div 12 =$	

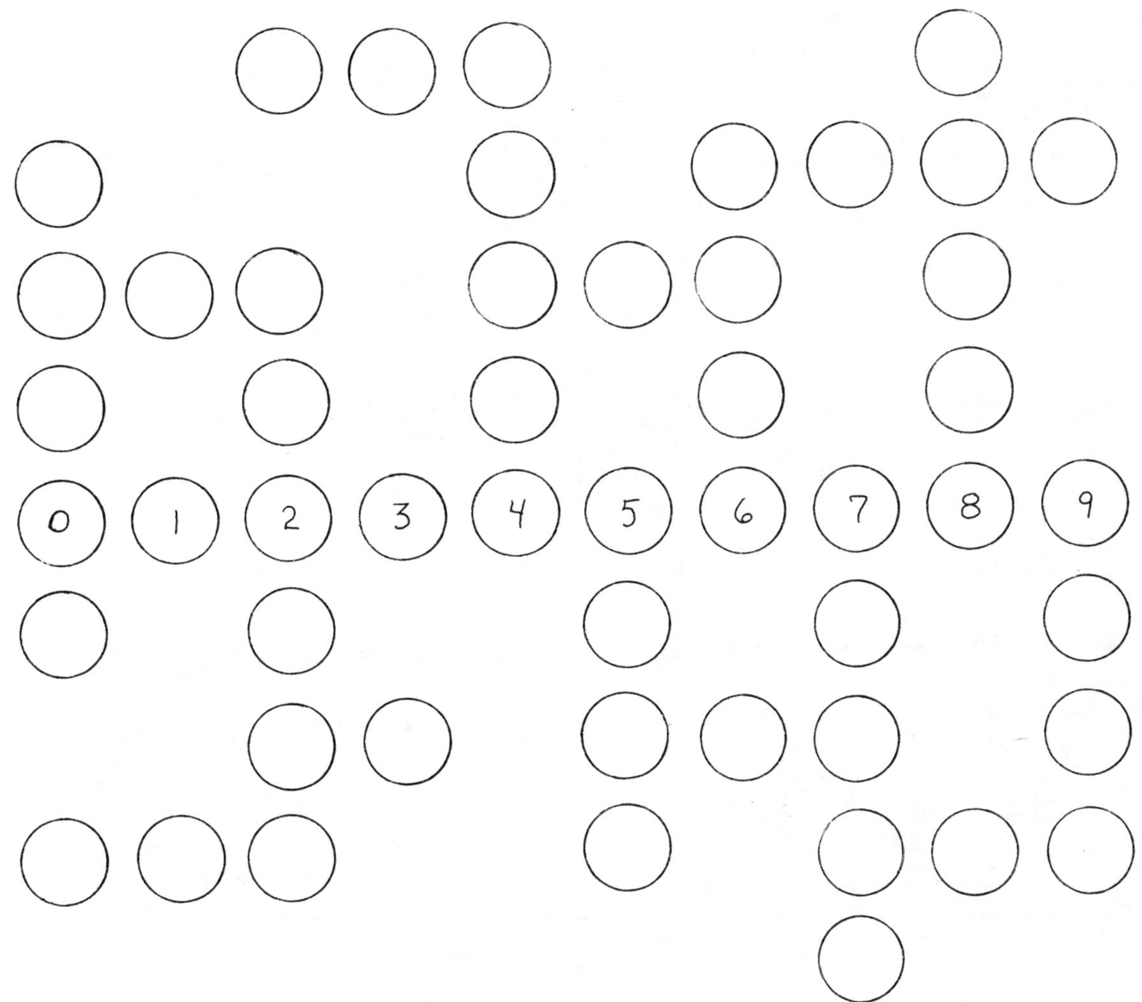

Solve all the problems below. Then fit each answer into the circlegram.

<u>Across</u>

```
 369        892         25
+605       -647       x  3


3 )1236    6 )2208


436        128        110
x 4       +235       +106
          +217       +127
```

<u>Down</u>

```
 265       -55197      -41147
 384       -40539      -16439
+972
+195


8262       14043       -13247
x  9       x  18       - 3897


5 )27160            7 )366548
```

Cut out the squares. Fit them together so that the touching edges name the same number.

193 - 78 3 11x11 56÷4	14 5x7 7 82	810 ÷ 90 314 86x7 210	18 154 13x3 12x12	87 10 730÷10 57+92
152 13x11 591÷3 27x14	102 73 639÷9 293-57	478 89 297÷9 827+956	225 400÷200 417 426+397	59x3 99 72÷6 29x3
492÷6 12 10x8 17x6	823 18x18 28÷7 161+259	983-495 87 8x9 72÷4	378 33 61x4 6x5	15x14 9x9 143 86+392
8x8 121 57 440	257+183 56÷8 93 15x8	149 197 138 125÷5	120 80 4x3 595÷7	93 50 35 177
85 71 9x17 15x15	24x83 602 500÷50 19x8	419 39 200÷4 16x8	562-493 72 84÷3 273+146	54 876÷2 324 817-534
813+729 28 57÷19 279÷3	25 244 438 413-216	144 6x10 47 9	236 46x3 2 6x9	128 94÷2 9x11 1992

Connect the answers in order.

1. 27 + 31 + 24 + 16 = _______
2. 258 ÷ 3 = _______
3. 37 x 14 – 500 = _______
4. 17 + 29 + 13 + 38 = _______
5. 593 – 497 = _______
6. 564 ÷ 6 – 90 = _______
7. 813 + 746 – 1550 = _______
8. 92 + 431 – 269 – 248 = _______
9. 32 x 29 ÷ 464 = _______
10. 630 ÷ 15 = _______
11. 570 – 439 – 128 = _______
12. 1702 – 1691 = _______
13. 148 x 6 ÷ 111 = _______
14. 2175 ÷ 25 = _______
15. 125 + 327 – 432 – 15 = _______
16. 18 x 47 – 845 = _______

17. 2576 – 2488 – 78 = _______
18. 56 + 173 + 9 – 226 = _______
19. 287 x 3 – 854 = _______
20. 3212 ÷ 44 = _______
21. 7 x 7 x 2 = _______
22. 273 + 56 – 300 = _______
23. 9 x 7 x 5 – 300 = _______
24. 455 ÷ 5 = _______
25. 263 – 192 – 50 = _______
26. 53 x 15 – 770 = _______
27. 280 ÷ 10 = _______
28. 1936 – 928 – 939 = _______
29. 4300 ÷ 100 = _______
30. 563 + 283 – 800 = _______
31. 29 x 15 – 338 = _______

Find the shortest path from entrance to exit by finding fractions equivalent to the fraction $\frac{2}{3}$.

$\frac{2}{3}$ Exit

$\frac{28}{42}$

$\frac{30}{40}$

$\frac{40}{60}$

$\frac{30}{50}$

$\frac{4}{12}$

$\frac{4}{9}$

$\frac{26}{39}$

$\frac{10}{16}$

$\frac{4}{9}$

$\frac{21}{30}$

$\frac{16}{24}$

$\frac{30}{45}$

$\frac{18}{27}$

$\frac{18}{24}$

$\frac{14}{20}$

$\frac{22}{33}$

$\frac{8}{12}$

$\frac{12}{16}$

$\frac{24}{36}$

$\frac{6}{8}$

$\frac{12}{18}$

$\frac{12}{15}$

$\frac{4}{12}$

$\frac{6}{12}$

$\frac{1}{3}$

$\frac{9}{12}$

$\frac{8}{10}$

$\frac{1}{2}$

$\frac{14}{21}$

$\frac{3}{7}$

$\frac{10}{12}$

$\frac{10}{15}$

$\frac{10}{16}$

$\frac{20}{30}$

$\frac{3}{7}$

$\frac{2}{5}$

$\frac{5}{10}$

$\frac{5}{6}$

$\frac{4}{6}$

$\frac{4}{5}$

$\frac{6}{9}$

$\frac{5}{15}$

22

$\frac{2}{3}$ Entrance

ADDITION OF FRACTIONS

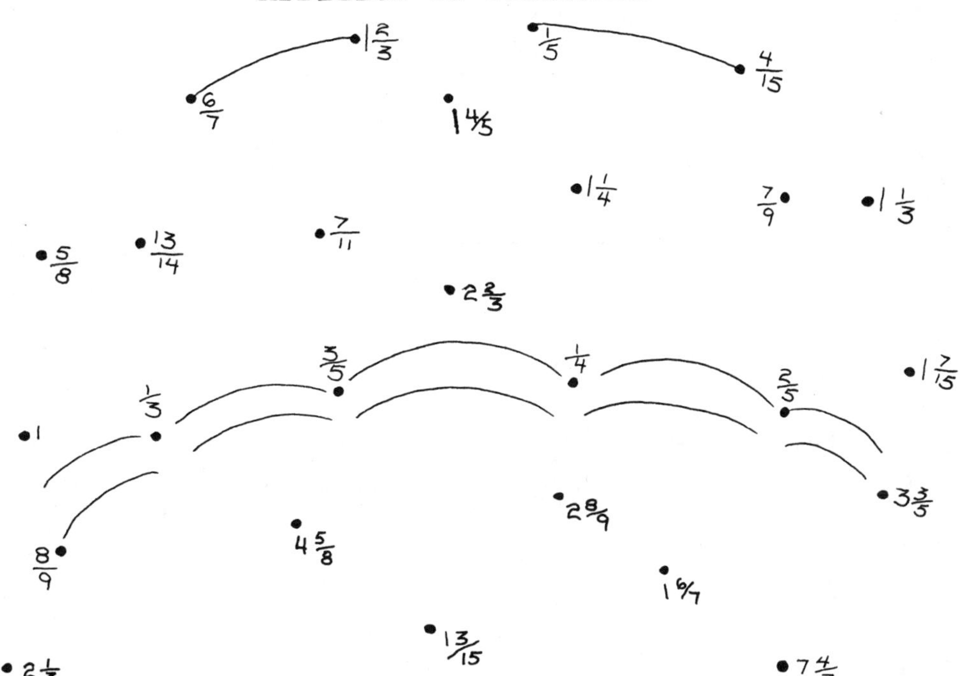

Connect the answers in order.

1. $\dfrac{1}{5} + \dfrac{3}{5} =$ ______

2. $\dfrac{1}{3} + \dfrac{1}{3} =$ ______

3. $\dfrac{4}{9} + \dfrac{4}{9} =$ ______

4. $\dfrac{1}{2} + \dfrac{1}{2} =$ ______

5. $\dfrac{3}{8} + \dfrac{2}{8} =$ ______

6. $\dfrac{4}{7} + \dfrac{2}{7} =$ ______

7. $\dfrac{6}{14} + \dfrac{7}{14} =$ ______

8. $\dfrac{1}{6} + \dfrac{1}{6} =$ ______

9. $\dfrac{3}{4} + \dfrac{3}{4} =$ ______

10. $\dfrac{5}{12} + \dfrac{1}{12} =$ ______

11. $\dfrac{3}{10} + \dfrac{3}{10} =$ ______

12. $\dfrac{5}{11} + \dfrac{2}{11} =$ ______

13. $\dfrac{5}{6} + \dfrac{5}{6} =$ ______

14. $\dfrac{1}{15} + \dfrac{2}{15} =$ ______

15. $\dfrac{3}{8} + \dfrac{7}{8} =$ ______

16. $\dfrac{1}{16} + \dfrac{3}{16} =$ ______

17. $\dfrac{4}{5} + \dfrac{3}{5} =$ ______

18. $\dfrac{1}{9} + \dfrac{4}{9} =$ ______

19. $\dfrac{3}{20} + \dfrac{5}{20} =$ ______

20. $\dfrac{7}{18} + \dfrac{7}{18} =$ ______

21. $\dfrac{1}{30} + \dfrac{7}{30} =$ ______

22. $\dfrac{5}{12} + \dfrac{11}{12} =$ ______

23. $\dfrac{14}{15} + \dfrac{8}{15} =$ ______

24. $1\dfrac{2}{5} + 2\dfrac{1}{5} =$ ______

25. $4\dfrac{2}{7} + 3\dfrac{2}{7} =$ ______

26. $5\dfrac{1}{2} + 3\dfrac{1}{2} =$ ______

27. $2\dfrac{1}{4} + 3\dfrac{1}{4} =$ ______

28. $4\dfrac{5}{6} + 2\dfrac{1}{6} =$ ______

29. $4\dfrac{3}{4} + 1\dfrac{3}{4} =$ ______

30. $1\dfrac{5}{8} + 1\dfrac{7}{8} =$ ______

31. $2\dfrac{3}{4} + 5 =$ ______

32. $3\dfrac{7}{9} + 1 =$ ______

33. $2\dfrac{6}{7} + 1\dfrac{1}{7} =$ ______

ADDITION OF FRACTIONS

Martin Luther King was a famous

$$\overline{13}\ \overline{5}\ \overline{6}\ \overline{5}\ \overline{23} \qquad \overline{11}\ \overline{5}\ \overline{3}\ \overline{4}\ \overline{7}\ \overline{8} \qquad \overline{23}\ \overline{12}\ \overline{2}\ \overline{10}\ \overline{12}\ \overline{11}$$

Solve the problems. Find the number that each letter represents.
Fill in the blank with the letters each number represents.

1. $\dfrac{1}{3} + \dfrac{1}{6} = \dfrac{1}{A}$

2. $\dfrac{1}{2} + \dfrac{1}{4} = \dfrac{3}{H}$

3. $\dfrac{1}{8} + \dfrac{3}{4} = \dfrac{7}{S}$

4. $\dfrac{1}{4} + \dfrac{1}{6} = \dfrac{I}{12}$

5. $\dfrac{2}{5} + \dfrac{3}{10} = \dfrac{7}{D}$

6. $\dfrac{2}{9} + \dfrac{1}{6} = \dfrac{T}{18}$

7. $\dfrac{5}{12} + \dfrac{1}{2} = \dfrac{11}{E}$

8. $\dfrac{1}{6} + \dfrac{3}{8} = \dfrac{C}{24}$

9. $\dfrac{1}{4} + \dfrac{3}{10} = \dfrac{R}{20}$

10. $\dfrac{5}{12} + \dfrac{1}{4} = \dfrac{2}{G}$

11. $\dfrac{1}{6} + \dfrac{3}{5} = \dfrac{L}{30}$

12. $\dfrac{3}{10} + \dfrac{8}{15} = \dfrac{5}{V}$

ADDITION OF FRACTIONS

Use three of these fractions to make each of the
number sentences true.

$\frac{1}{2}$ $\frac{1}{3}$ $\frac{2}{3}$ $\frac{1}{4}$ $\frac{1}{6}$ $\frac{1}{5}$

1. $\bigcirc + \bigcirc + \bigcirc = 1\frac{5}{12}$

2. $\bigcirc + \bigcirc + \bigcirc = 1$

3. $\bigcirc + \bigcirc + \bigcirc = 1\frac{1}{4}$

4. $\bigcirc + \bigcirc + \bigcirc = 1\frac{1}{2}$

5. $\bigcirc + \bigcirc + \bigcirc = \frac{7}{10}$

6. $\bigcirc + \bigcirc + \bigcirc = 1\frac{11}{30}$

7. $\bigcirc + \bigcirc + \bigcirc = \frac{3}{4}$

8. $\bigcirc + \bigcirc + \bigcirc = \frac{47}{60}$

9. $\bigcirc + \bigcirc + \bigcirc = 1\frac{1}{12}$

10. $\bigcirc + \bigcirc + \bigcirc = 1\frac{1}{30}$

SUBTRACTION OF FRACTIONS

$\frac{3}{4}$ •   • $\frac{1}{7}$

$\frac{1}{4}$ •   • $\frac{4}{5}$

$\frac{1}{2}$ •   • $\frac{2}{3}$

$\frac{1}{3}$ •   • $\frac{1}{11}$

1 •   • $\frac{3}{5}$

Connect "a" and "b" answers for each problem. Then cut along the
outside lines of the figure formed, fold along the inside lines,
and tape the figure together to form a cube.

	(a)		(b)
1. $\frac{2}{3} - \frac{1}{3} =$	_____	$\frac{5}{11} - \frac{4}{11} =$	_____
2. $\frac{4}{7} - \frac{3}{7} =$	_____	$\frac{4}{5} - \frac{1}{5} =$	_____
3. $\frac{3}{4} - \frac{1}{4} =$	_____	$\frac{7}{9} - \frac{1}{9} =$	_____
4. $\frac{3}{2} - \frac{1}{2} =$	_____	$\frac{9}{10} - \frac{3}{10} =$	_____
5. $\frac{11}{16} - \frac{3}{16} =$	_____	$\frac{5}{8} - \frac{3}{8} =$	_____
6. $\frac{13}{14} - \frac{11}{14} =$	_____	$\frac{11}{12} - \frac{2}{12} =$	_____
7. $\frac{5}{6} - \frac{1}{6} =$	_____	$\frac{14}{15} - \frac{2}{15} =$	_____
8. $\frac{17}{20} - \frac{12}{20} =$	_____	$\frac{23}{25} - \frac{3}{25} =$	_____
9. $\frac{9}{7} - \frac{2}{7} =$	_____	$\frac{23}{24} - \frac{5}{24} =$	_____

SUBTRACTION OF FRACTIONS

Connect the answers in order.

1. $\dfrac{8}{9} - \dfrac{1}{9} =$ _____

2. $\dfrac{6}{8} - \dfrac{3}{8} =$ _____

3. $\dfrac{3}{5} - \dfrac{1}{5} =$ _____

4. $\dfrac{6}{7} - \dfrac{1}{7} =$ _____

5. $\dfrac{9}{20} - \dfrac{4}{20} =$ _____

6. $\dfrac{7}{21} - \dfrac{1}{21} =$ _____

7. $\dfrac{7}{15} - \dfrac{4}{15} =$ _____

8. $\dfrac{9}{12} - \dfrac{8}{12} =$ _____

9. $\dfrac{7}{26} - \dfrac{4}{26} =$ _____

10. $\dfrac{1}{4} - \dfrac{1}{8} =$ _____

11. $\dfrac{1}{2} - \dfrac{1}{6} =$ _____

12. $\dfrac{2}{3} - \dfrac{2}{9} =$ _____

13. $\dfrac{7}{8} - \dfrac{1}{4} =$ _____

14. $\dfrac{3}{4} - \dfrac{1}{12} =$ _____

15. $\dfrac{2}{3} - \dfrac{1}{6} =$ _____

16. $\dfrac{11}{12} - \dfrac{1}{2} =$ _____

17. $\dfrac{9}{10} - \dfrac{1}{5} =$ _____

18. $\dfrac{3}{4} - \dfrac{1}{6} =$ _____

19. $\dfrac{1}{2} - \dfrac{1}{3} =$ _____

20. $\dfrac{5}{4} - \dfrac{1}{3} =$ _____

21. $\dfrac{5}{6} - \dfrac{4}{9} =$ _____

22. $\dfrac{3}{4} - \dfrac{3}{5} =$ _____

23. $\dfrac{7}{8} - \dfrac{5}{6} =$ _____

24. $\dfrac{7}{12} - \dfrac{3}{8} =$ _____

25. $\dfrac{2}{3} - \dfrac{3}{5} =$ _____

26. $\dfrac{5}{7} - \dfrac{1}{2} =$ _____

27. $\dfrac{4}{9} - \dfrac{5}{12} =$ _____

28. $\dfrac{11}{16} - \dfrac{1}{4} =$ _____

29. $\dfrac{3}{4} - \dfrac{5}{9} =$ _____

30. $\dfrac{5}{14} - \dfrac{2}{7} =$ _____

31. $\dfrac{5}{8} - \dfrac{1}{3} =$ _____

32. $\dfrac{9}{10} - \dfrac{7}{15} =$ _____

33. $\dfrac{9}{20} - \dfrac{3}{25} =$ _____

34. $\dfrac{3}{2} - \dfrac{1}{2} =$ _____

35. $\dfrac{13}{3} - \dfrac{1}{3} =$ _____

36. $\dfrac{19}{5} - \dfrac{4}{5} =$ _____

37. $\dfrac{2}{3} - \dfrac{2}{3} =$ _____

38. $\dfrac{93}{10} - \dfrac{3}{10} =$ _____

39. $\dfrac{13}{2} - \dfrac{1}{2} =$ _____

40. $\dfrac{35}{4} - \dfrac{7}{4} =$ _____

A winner of two Nobel Prizes:

$$\overline{19}\ \overline{6}\ \overline{11}\ \overline{4}\ \overline{13} \qquad \overline{1}\ \overline{7}\ \overline{10}\ \overline{24}\ \overline{17}\ \overline{24}\ \overline{18}\ \overline{1}\ \overline{7}\ \overline{6} \qquad \overline{5}\ \overline{3}\ \overline{11}\ \overline{4}\ \overline{13}$$

Solve the problems. Find the number that each letter represents.
Fill in the blank with the letters each number represents.

1. $\dfrac{4}{5} - \dfrac{1}{2} = \dfrac{3}{L}$

2. $\dfrac{2}{9} - \dfrac{1}{6} = \dfrac{1}{W}$

3. $\dfrac{5}{12} - \dfrac{1}{4} = \dfrac{1}{A}.$

4. $\dfrac{7}{8} - \dfrac{1}{6} = \dfrac{D}{24}$

5. $\dfrac{9}{10} - \dfrac{7}{15} = \dfrac{E}{30}$

6. $\dfrac{11}{18} - \dfrac{5}{12} = \dfrac{K}{36}$

7. $\dfrac{9}{20} - \dfrac{4}{15} = \dfrac{R}{60}$

8. $10\dfrac{4}{5} - 4\dfrac{7}{15} = 6\dfrac{1}{U}$

9. $9\dfrac{2}{3} - 3\dfrac{5}{8} = 6\dfrac{S}{24}$

10. $2\dfrac{14}{25} - 1\dfrac{2}{5} = 1\dfrac{I}{25}$

11. $3\dfrac{7}{12} - 2\dfrac{3}{8} = 1\dfrac{5}{O}$

12. $4\dfrac{1}{6} - 2\dfrac{3}{8} = 1\dfrac{M}{24}$

13. $8\dfrac{1}{2} - 5\dfrac{2}{3} = 2\dfrac{C}{6}$

Founder of Modern Astronomy:

$$\overline{9}\ \overline{5}\ \overline{22}\ \overline{4}\ \overline{24}\ \overline{3}\ \overline{8}\ \overline{1} \qquad \overline{22}\ \overline{24}\ \overline{7}\ \overline{10}\ \overline{2}\ \overline{9}\ \overline{5}\ \overline{22}\ \overline{14}\ \overline{1}$$

Solve the problems. Find the number that each letter represents.
Fill in the blanks with the letters each number represents.

1. $\frac{1}{3} \times 9 = \dfrac{}{\overline{L}}$

2. $12 \times \frac{1}{6} = \dfrac{}{\overline{R}}$

3. $\frac{1}{3} \times \frac{4}{5} = \dfrac{H}{15}$

4. $\frac{4}{15} \times \frac{3}{8} = \dfrac{S}{10}$

5. $\frac{10}{21} \times \frac{7}{12} = \dfrac{I}{18}$

6. $\frac{4}{7} \times \frac{5}{6} = \dfrac{E}{21}$

7. $\frac{16}{25} \times \frac{15}{18} = \dfrac{A}{15}$

8. $\frac{5}{22} \times \frac{11}{12} = \dfrac{5}{O}$

9. $\frac{7}{9} \times \frac{6}{11} = \dfrac{U}{33}$

10. $\frac{2}{3} \times \frac{5}{6} = \dfrac{5}{N}$

11. $\frac{8}{9} \times \frac{11}{12} = \dfrac{C}{27}$

12. $\frac{14}{27} \times \frac{9}{16} = \dfrac{P}{24}$

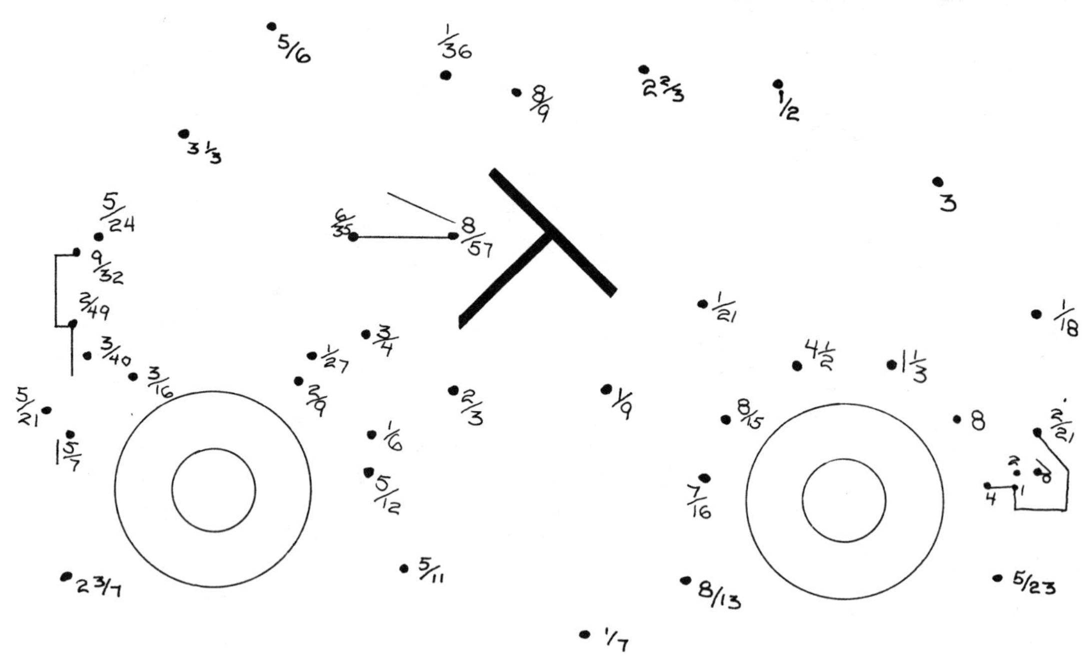

Connect the answers in order.

1. $6 \times \frac{2}{3} =$ _____

2. $\frac{4}{5} \times 10 =$ _____

3. $12 \times \frac{1}{9} =$ _____

4. $15 \times \frac{3}{10} =$ _____

5. $\frac{2}{3} \times \frac{4}{5} =$ _____

6. $\frac{3}{8} \times \frac{7}{6} =$ _____

7. $\frac{5}{9} \times \frac{3}{4} =$ _____

8. $\frac{5}{12} \times \frac{8}{15} =$ _____

9. $\frac{3}{8} \times \frac{1}{2} =$ _____

10. $\frac{8}{7} \times \frac{3}{2} =$ _____

11. $\frac{2}{7} \times \frac{5}{6} =$ _____

12. $\frac{15}{16} \times \frac{2}{25} =$ _____

13. $\frac{2}{9} \times \frac{1}{6} =$ _____

14. $\frac{3}{8} \times \frac{4}{9} =$ _____

15. $\frac{9}{10} \times \frac{5}{6} =$ _____

16. $\frac{4}{21} \times \frac{3}{14} =$ _____

17. $\frac{9}{20} \times \frac{5}{8} =$ _____

18. $\frac{5}{6} \times \frac{1}{4} =$ _____

19. $\frac{10}{21} \times \frac{9}{25} =$ _____

20. $\frac{3}{8} \times \frac{2}{27} =$ _____

21. $\frac{20}{21} \times \frac{14}{15} =$ _____

22. $\frac{4}{15} \times \frac{10}{19} =$ _____

23. $\frac{7}{9} \times \frac{6}{7} =$ _____

24. $\frac{4}{5} \times \frac{2}{9} \times \frac{5}{8} =$ _____

25. $\frac{1}{6} \times \frac{8}{21} \times \frac{3}{4} =$ _____

26. $\frac{5}{6} \times \frac{3}{10} \times \frac{2}{9} =$ _____

27. $\frac{10}{27} \times \frac{3}{4} \times \frac{12}{35} =$ _____

28. $\frac{2}{3} \times 0 \times \frac{5}{8} =$ _____

29. $26 \times \frac{1}{13} =$ _____

30. $\frac{14}{17} \times \frac{17}{14} =$ _____

MULTIPLICATION OF FRACTIONS

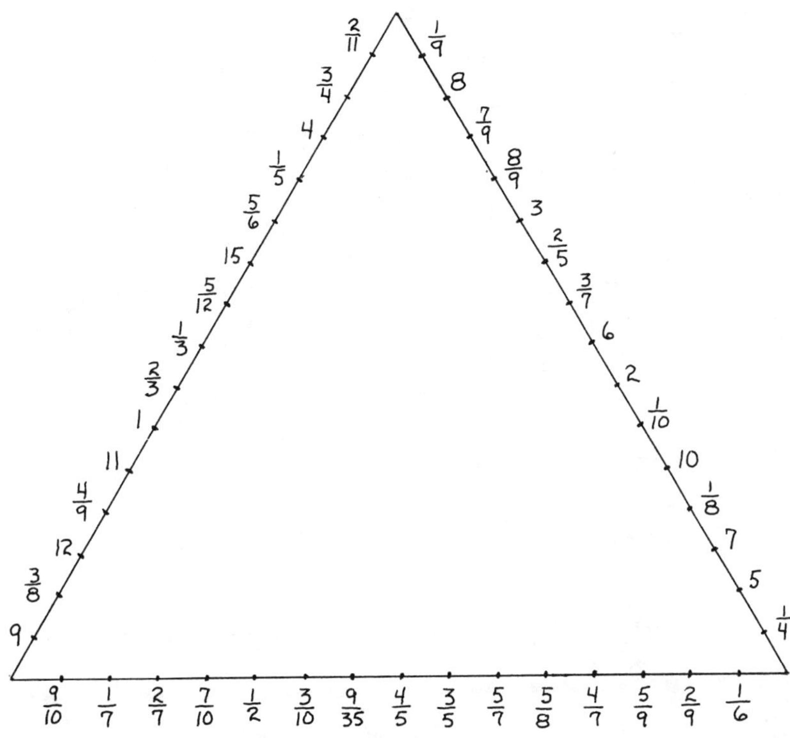

Connect "a" and "b" answers for each of the following problems.

	(a)		(b)		(a)		(b)
1. $\frac{1}{2} \times \frac{4}{5} =$	____	$6 \times \frac{2}{3} =$	____	13. $\frac{3}{5} \times 10 =$	____	$\frac{4}{9} \times \frac{3}{8} =$	____
2. $\frac{2}{3} \times \frac{2}{3} =$	____	$\frac{4}{7} \times \frac{7}{8} =$	____	14. $\frac{1}{12} \times 9 =$	____	$2 \times \frac{3}{14} =$	____
3. $12 \times \frac{1}{6} =$	____	$\frac{5}{9} \times \frac{6}{15} =$	____	15. $\frac{5}{8} \times \frac{8}{9} =$	____	$\frac{7}{30} \times \frac{3}{7} =$	____
4. $\frac{3}{10} \times \frac{6}{7} =$	____	$\frac{15}{22} \times \frac{11}{20} =$	____	16. $\frac{3}{7} \times \frac{7}{15} =$	____	$\frac{4}{7} \times 5\frac{1}{4} =$	____
5. $\frac{1}{3} \times \frac{1}{3} =$	____	$\frac{2}{5} \times \frac{5}{6} =$	____	17. $\frac{3}{5} \times \frac{20}{21} =$	____	$20 \times \frac{1}{2} =$	____
6. $15 \times \frac{1}{3} =$	____	$\frac{1}{5} \times 3 =$	____	18. $\frac{3}{4} \times \frac{5}{9} =$	____	$6\frac{2}{5} \times 1\frac{1}{4} =$	____
7. $\frac{2}{13} \times \frac{13}{14} =$	____	$\frac{8}{9} \times \frac{3}{4} =$	____	19. $\frac{2}{5} \times \frac{3}{4} =$	____	$18 \times \frac{2}{3} =$	____
8. $\frac{2}{9} \times \frac{7}{2} =$	____	$30 \times \frac{1}{2} =$	____	20. $9\frac{1}{3} \times \frac{3}{4} =$	____	$\frac{1}{7} \times 5 =$	____
9. $\frac{5}{6} \times \frac{3}{20} =$	____	$\frac{5}{4} \times \frac{1}{2} =$	____	21. $\frac{4}{5} \times \frac{7}{8} =$	____	$33 \times \frac{1}{3} =$	____
10. $16 \times \frac{1}{18} =$	____	$\frac{10}{11} \times \frac{11}{12} =$	____	22. $1\frac{1}{3} \times \frac{3}{5} =$	____	$6 \times 1\frac{1}{2} =$	____
11. $\frac{2}{3} \times 1\frac{1}{2} =$	____	$\frac{4}{6} \times \frac{3}{7} =$	____	23. $\frac{3}{5} \times \frac{5}{9} =$	____	$\frac{1}{30} \times 27 =$	____
12. $\frac{6}{7} \times \frac{14}{15} =$	____	$\frac{7}{10} \times \frac{5}{14} =$	____	24. $\frac{4}{5} \times \frac{5}{22} =$	____	$\frac{6}{7} \times 7 =$	____

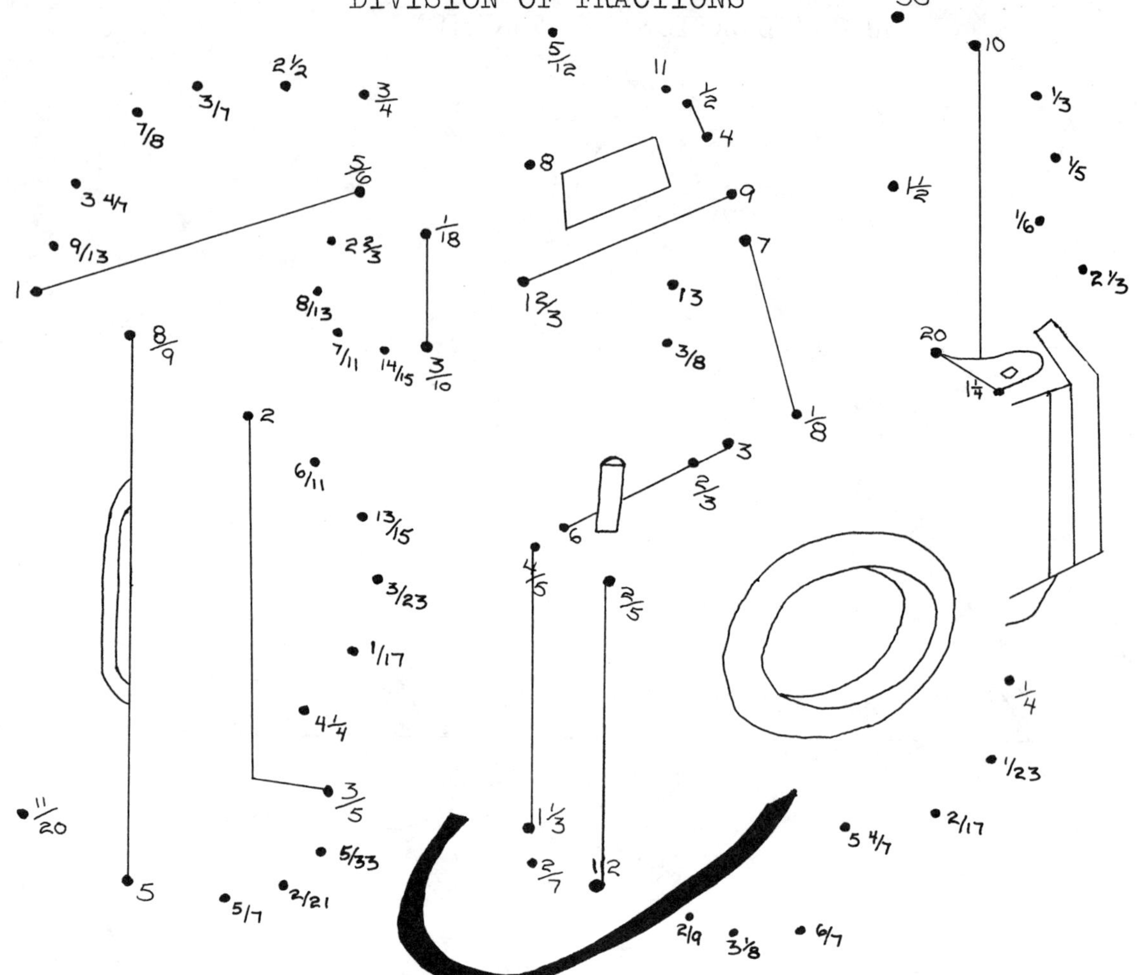

Connect the answers in order.

1. $\dfrac{1}{3} \div \dfrac{1}{2} =$ _______

2. $\dfrac{1}{5} \div \dfrac{2}{3} =$ _______

3. $\dfrac{4}{5} \div \dfrac{2}{5} =$ _______

4. $5 \div \dfrac{5}{6} =$ _______

5. $\dfrac{4}{7} \div \dfrac{5}{7} =$ _______

6. $\dfrac{1}{4} \div \dfrac{5}{8} =$ _______

7. $\dfrac{7}{8} \div \dfrac{7}{10} =$ _______

8. $\dfrac{1}{5} \div \dfrac{4}{5} =$ _______

9. $9 \div \dfrac{3}{4} =$ _______

10. $\dfrac{4}{7} \div 2 =$ _______

11. $\dfrac{2}{9} \div \dfrac{1}{6} =$ _______

12. $\dfrac{1}{2} \div \dfrac{5}{6} =$ _______

13. $\dfrac{5}{7} \div \dfrac{1}{7} =$ _______

14. $\dfrac{3}{10} \div \dfrac{6}{11} =$ _______

15. $\dfrac{12}{14} \div \dfrac{6}{7} =$ _______

16. $\dfrac{8}{15} \div \dfrac{3}{5} =$ _______

17. $\dfrac{4}{9} \div 8 =$ _______

18. $\dfrac{5}{8} \div \dfrac{3}{4} =$ _______

19. $\dfrac{9}{13} \div \dfrac{12}{13} =$ _______

20. $\dfrac{1}{9} \div \dfrac{4}{15} =$ _______

21. $\dfrac{2}{5} \div \dfrac{12}{15} =$ _______

22. $\dfrac{8}{9} \div \dfrac{1}{9} =$ _______

23. $\dfrac{5}{6} \div \dfrac{1}{2} =$ _______

24. $\dfrac{1}{4} \div \dfrac{1}{12} =$ _______

25. $\dfrac{5}{4} \div 10 =$ _______

26. $18 \div \dfrac{9}{10} =$ _______

27. $\dfrac{4}{15} \div \dfrac{8}{45} =$ _______

28. $\dfrac{7}{17} \div \dfrac{1}{17} =$ _______

29. $\dfrac{9}{10} \div \dfrac{1}{10} =$ _______

30. $\dfrac{2}{3} \div \dfrac{1}{6} =$ _______

31. $2 \div \dfrac{1}{5} =$ _______

32. $\dfrac{1}{2} \div \dfrac{1}{60} =$ _______

33. $1 \div \dfrac{1}{11} =$ _______

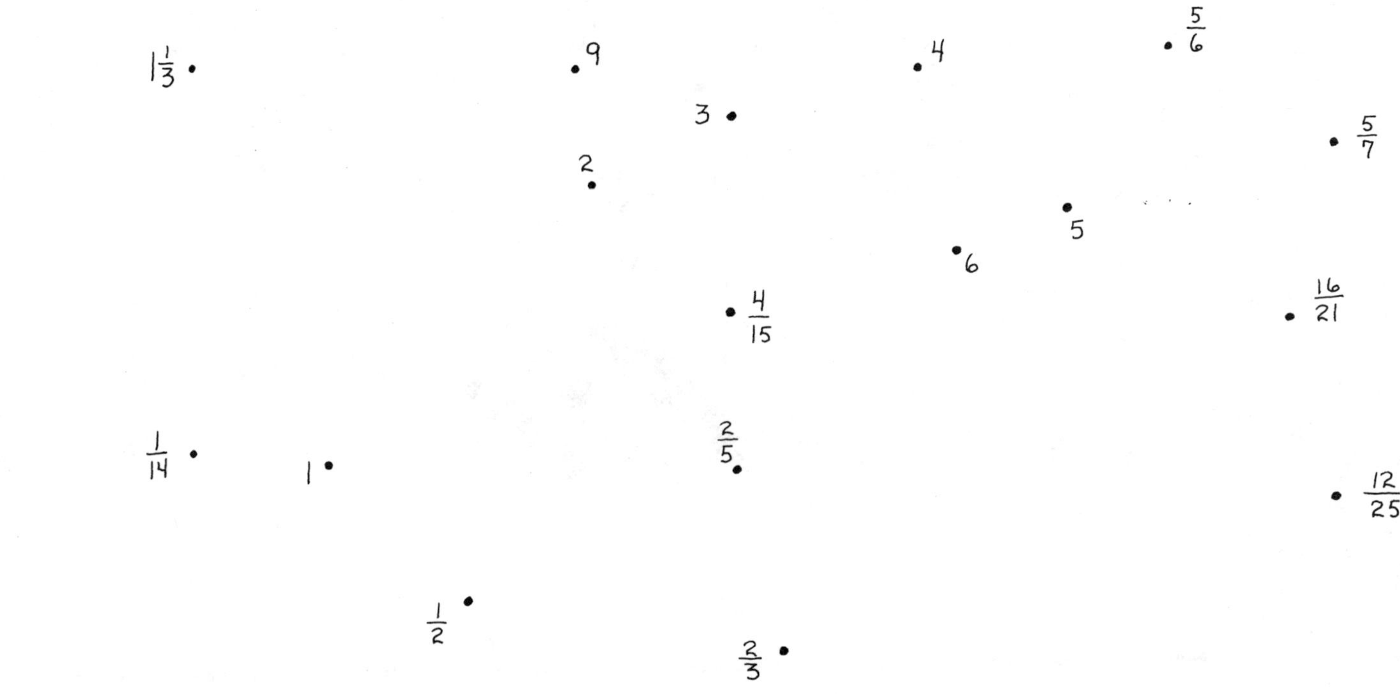

Connect "a" and "b" answers for each problem. Then cut out the polygons formed and put them together to form a triangle.

	(a)	(b)		(a)	(b)
1. $\frac{1}{3} \div \frac{1}{2} =$	_____	$\frac{2}{5} \div \frac{5}{6} =$ _____	10. $\frac{3}{7} \div \frac{1}{21} =$	_____	$\frac{1}{6} \div \frac{1}{8} =$ _____
2. $\frac{2}{3} \div \frac{4}{5} =$	_____	$\frac{3}{7} \div \frac{3}{5} =$ _____	11. $\frac{12}{7} \div \frac{4}{7} =$	_____	$\frac{8}{9} \div \frac{4}{27} =$ _____
3. $\frac{2}{5} \div \frac{1}{5} =$	_____	$\frac{2}{9} \div \frac{5}{6} =$ _____	12. $\frac{2}{3} \div \frac{5}{3} =$	_____	$\frac{3}{4} \div \frac{9}{8} =$ _____
4. $\frac{15}{8} \div \frac{5}{8} =$	_____	$3 \div \frac{3}{4} =$ _____	13. $\frac{2}{7} \div \frac{4}{7} =$	_____	$\frac{2}{5} \div \frac{3}{2} =$ _____
5. $\frac{2}{3} \div \frac{1}{2} =$	_____	$\frac{5}{7} \div 10 =$ _____	14. $\frac{2}{4} \div \frac{3}{6} =$	_____	$\frac{4}{7} \div \frac{2}{7} =$ _____
6. $\frac{4}{7} \div \frac{4}{7} =$	_____	$\frac{8}{12} \div \frac{4}{3} =$ _____	15. $\frac{1}{2} \div 7 =$	_____	$6 \div \frac{2}{3} =$ _____
7. $\frac{2}{11} \div \frac{5}{11} =$	_____	$\frac{4}{9} \div \frac{7}{12} =$ _____	16. $1 \div \frac{1}{4} =$	_____	$\frac{4}{5} \div \frac{2}{15} =$ _____
8. $\frac{5}{6} \div \frac{3}{18} =$	_____	$\frac{4}{7} \div \frac{4}{5} =$ _____	17. $\frac{1}{2} \div \frac{3}{5} =$	_____	$\frac{5}{6} \div \frac{1}{6} =$ _____
9. $\frac{4}{3} \div \frac{7}{4} =$	_____	$\frac{3}{5} \div \frac{5}{4} =$ _____			

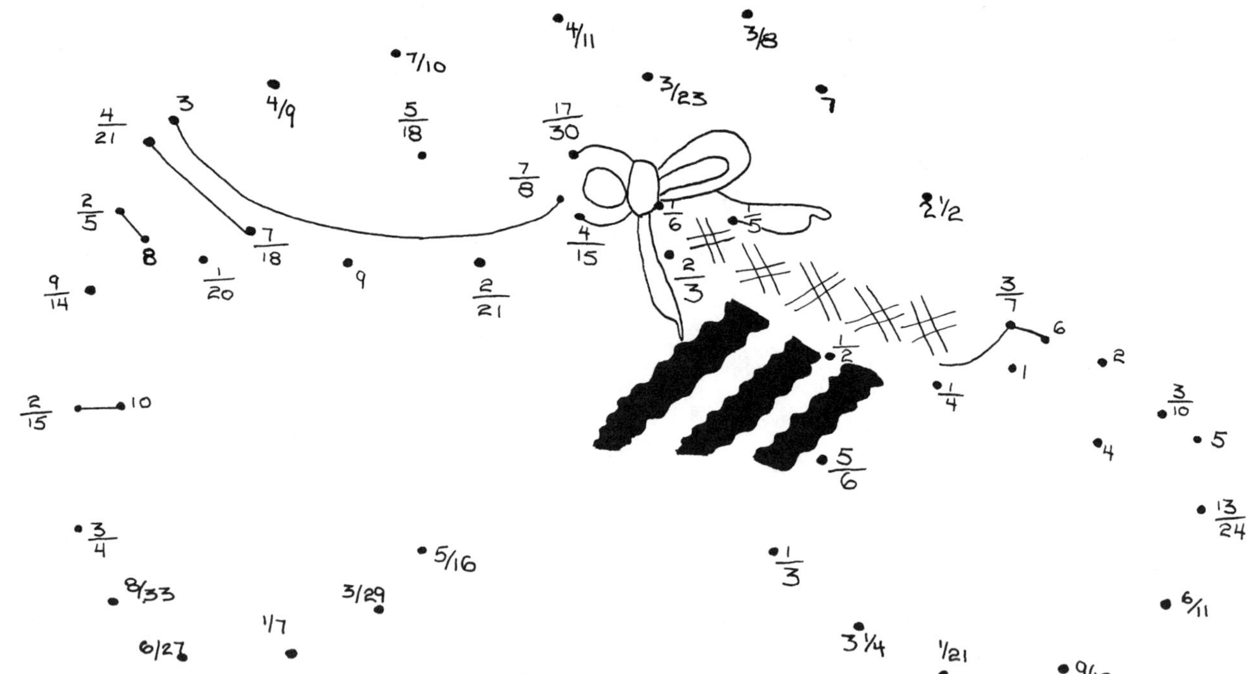

FRACTION OPERATIONS

Connect the answers in order.

1. $\frac{2}{7} + \frac{1}{7} =$ ____

2. $\frac{4}{5} - \frac{3}{5} =$ ____

3. $\frac{1}{3} \times \frac{1}{2} =$ ____

4. $\frac{2}{9} \div \frac{1}{3} =$ ____

5. $\frac{5}{6} - \frac{2}{6} =$ ____

6. $\frac{1}{2} \times \frac{1}{2} =$ ____

7. $\frac{4}{9} + \frac{5}{9} =$ ____

8. $2 \div \frac{1}{3} =$ ____

9. $\frac{4}{5} \div \frac{2}{5} =$ ____

10. $\frac{9}{10} - \frac{3}{5} =$ ____

11. $12 \times \frac{1}{3} =$ ____

12. $\frac{1}{2} + \frac{1}{3} =$ ____

13. $5 \div \frac{1}{2} =$ ____

14. $\frac{2}{5} \times 20 =$ ____

15. $\frac{4}{5} - \frac{3}{4} =$ ____

16. $\frac{2}{9} + \frac{1}{6} =$ ____

17. $6 \div \frac{2}{3} =$ ____

18. $\frac{2}{3} - \frac{4}{7} =$ ____

19. $\frac{2}{7} \times \frac{14}{15} =$ ____

20. $\frac{3}{8} + \frac{1}{2} =$ ____

21. $\frac{3}{10} + \frac{4}{15} =$ ____

22. $\frac{1}{2} - \frac{2}{9} =$ ____

23. $\frac{3}{7} \div \frac{1}{7} =$ ____

24. $\frac{2}{9} \times \frac{6}{7} =$ ____

25. $\frac{1}{7} \div \frac{5}{14} =$ ____

26. $\frac{2}{7} + \frac{5}{14} =$ ____

27. $\frac{8}{15} - \frac{2}{5} =$ ____

28. $\frac{9}{16} \div \frac{6}{8} =$ ____

29. $\frac{16}{21} \times \frac{7}{16} =$ ____

30. $\frac{3}{8} + \frac{1}{6} =$ ____

31. $\frac{5}{8} \div \frac{1}{8} =$ ____

32. $3 \times \frac{1}{10} =$ ____

FRACTION OPERATIONS

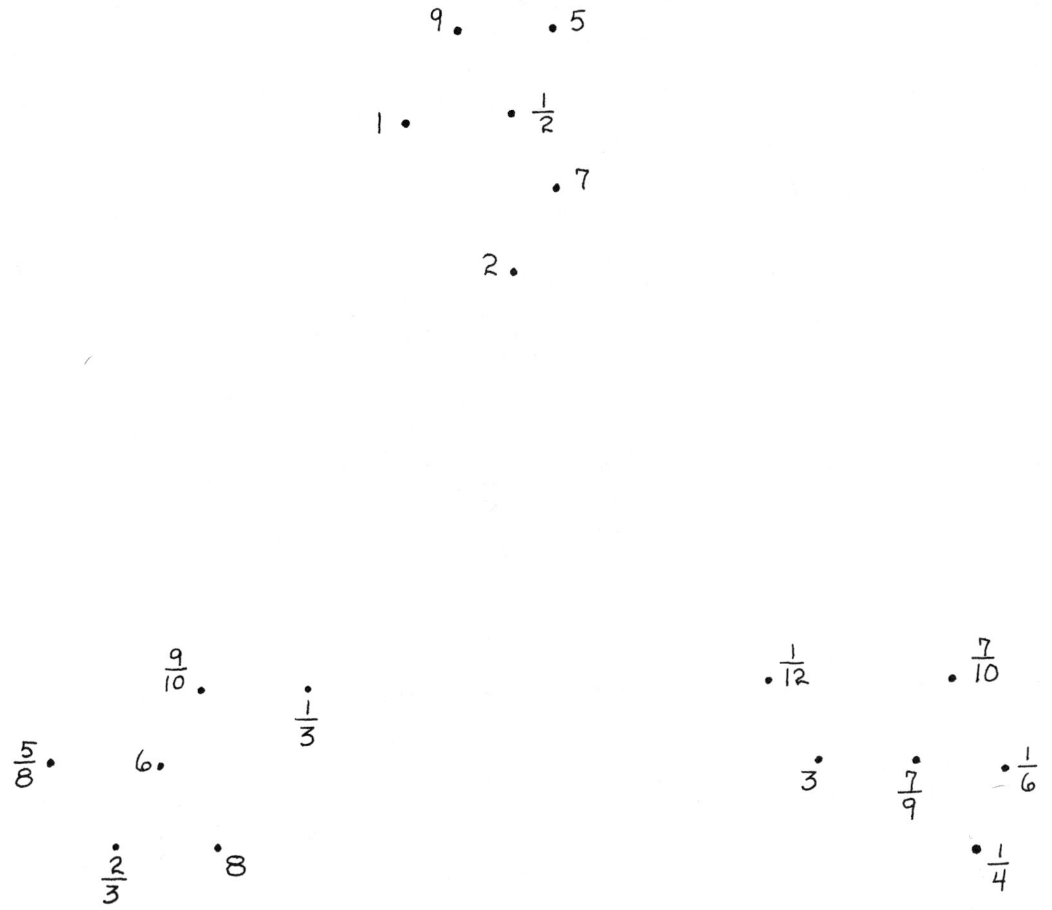
9
5
1
1/2
7
2
9/10
1/3
5/8
6
2/3
8
1/12
7/10
3
7/9
1/6
1/4

Connect "a" and "b" answer for each problem. Study the resulting figure.

	(a)		(b)			(a)		(b)	
1.	$\frac{2}{3} \times \frac{3}{4} =$	____	$\frac{4}{5} + \frac{1}{5} =$	____	9.	$\frac{1}{3} \times 15 =$	____	$\frac{6}{5} \div \frac{2}{15} =$	____
2.	$\frac{7}{8} - \frac{2}{8} =$	____	$\frac{2}{7} \div \frac{3}{7} =$	____	10.	$\frac{1}{2} - \frac{1}{3} =$	____	$\frac{5}{6} \div \frac{1}{6} =$	____
3.	$\frac{4}{9} + \frac{1}{3} =$	____	$\frac{2}{5} + \frac{3}{10} =$	____	11.	$\frac{2}{5} + \frac{1}{2} =$	____	$\frac{3}{4} - \frac{2}{3} =$	____
4.	$12 \times \frac{1}{6} =$	____	$\frac{3}{8} \div \frac{1}{8} =$	____	12.	$\frac{5}{9} \times \frac{3}{5} =$	____	$\frac{7}{6} \times 6 =$	____
5.	$\frac{11}{12} - \frac{3}{4} =$	____	$\frac{1}{2} - \frac{1}{4} =$	____	13.	$\frac{6}{11} \div \frac{1}{11} =$	____	$\frac{1}{3} \div \frac{3}{7} =$	____
6.	$1\frac{1}{2} \times 4 =$	____	$\frac{5}{7} \div \frac{1}{7} =$	____	14.	$\frac{3}{9} + \frac{1}{3} =$	____	$\frac{1}{12} \div \frac{1}{3} =$	____
7.	$\frac{9}{2} + \frac{3}{2} =$	____	$10 \times \frac{4}{5} =$	____	15.	$\frac{3}{4} \times 12 =$	____	$\frac{1}{4} + \frac{3}{8} =$	____
8.	$\frac{5}{6} - \frac{1}{3} =$	____	$\frac{1}{6} \div \frac{2}{3} =$	____					

READING AND WRITING DECIMALS

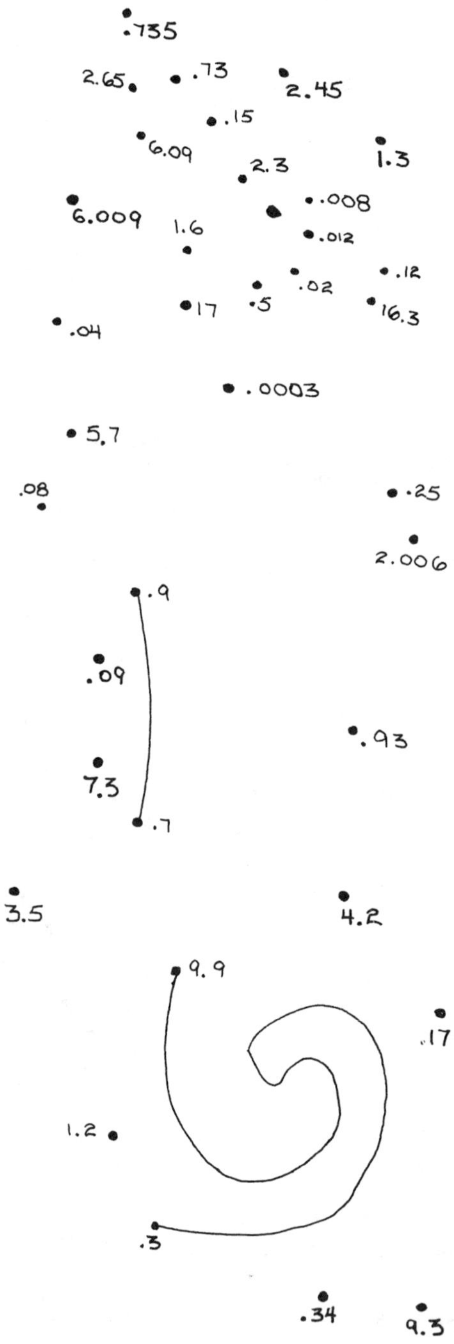

Connect the answers in order.

1. three tenths = ______
2. one and two tenths = ______
3. seven tenths = ______
4. three and five tenths = ______
5. eight hundredths = ______
6. nine tenths = ______
7. five and seven tenths = ______
8. four hundredths = ______
9. six and nine hundredths = ______
10. two and sixty-five hundredths = ______
11. seventy-three hundredths = ______
12. fifteen hundredths = ______
13. two and three tenths = ______
14. eight thousandths = ______
15. twelve thousandths = ______
16. twelve hundredths = ______
17. sixteen and three tenths = ______
18. two hundredths = ______
19. five tenths = ______
20. one and six tenths = ______
21. seventeen = ______
22. three ten-thousandths = ______
23. twenty-five hundredths = ______
24. two and six thousandths = ______
25. ninety-three hundredths = ______
26. nine and nine tenths = ______

FRACTIONS AND DECIMALS

Cut out the squares. Fit them together so that the touching
edges name the same number.

$\frac{17}{100}$ .003 .01 $\frac{3}{4}$	$\frac{1}{50}$ $\frac{1}{8}$ .89 .017	.53 $\frac{9}{10}$ $\frac{1}{6}$ .061	$\frac{17}{1000}$ $\frac{131}{1000}$ .2 .7	$\frac{5}{6}$ .09 $\frac{7}{50}$ .17
$\frac{17}{10000}$ $\frac{7}{100}$ .6 $\frac{3}{8}$	.875 $\frac{8}{11}$ .05 $\frac{2}{3}$	$\frac{7}{100}$ $\frac{1}{1000}$ .04 $\frac{2}{7}$	.4 $\frac{1}{10}$ $\frac{3}{1000}$ .02	.375 .423 $\frac{3}{10000}$ $\frac{4}{5}$
.3 $\frac{1}{5}$ $\frac{423}{1000}$ .07	.8 $\frac{1}{25}$ .29 .123	.543 $\frac{1}{9}$ $\frac{1}{4}$ $\frac{11}{100}$	$\frac{5}{8}$ $\frac{29}{100}$ $\frac{5}{7}$ .617	.75 $\frac{89}{100}$ .07 $\frac{3}{10}$
.63 .0003 $\frac{2}{9}$ .625	.11 $\frac{1}{3}$ .1 $\frac{97}{100}$	$\frac{7}{10}$ $\frac{1}{20}$ .001 $\frac{4}{9}$	.97 .432 .125 $\frac{73}{100}$	.92 .14 .5 $\frac{3}{50}$
.715 .25 $\frac{9}{100}$ $\frac{2}{5}$	$\frac{61}{1000}$ $\frac{3}{5}$ .34 $\frac{63}{100}$	.06 $\frac{1}{100}$ .9 .0017	$\frac{1}{12}$ $\frac{1}{2}$ .52 $\frac{53}{100}$	.73 .655 .131 $\frac{7}{8}$

ADDITION OF DECIMALS

Connect the answers in order.

1. .2 + .5 = _______
2. .3 + .6 = _______
3. .6 + .7 = _______
4. .5 + .9 = _______
5. .8 + .8 = _______
    STOP
6. 3.7 + .8 = _______
7. 8.4 + 6.9 = _______
8. 2.4 + 1.6 = _______
9. 3.9 + 5.7 = _______
10. 18.3 + 27.6 = _______
11. 59.8 + 57.2 = _______
12. .68 + .93 = _______
13. 3.21 + 2.79 = _______
14. 12.34 + 27.48 = _______
15. 5.6 + 12.34 = _______
16. 9.39 + 15.6 = _______
17. 28.7 + 4.64 = _______
18. .83 + 12.1 = _______
19. 9.7 + .56 = _______
20. 18.1 + .008 = _______
21. 5.613 + .52 = _______
22. .001 + .03 = _______
23. .023 + .14 = _______
24. 1.03 + .005 = _______
25. .2 + .3 + .5 = _______
26. .8 + .9 + .6 = _______
27. 2.3 + 15 = _______
28. 87 + .34 = _______
29. 962 + 1.3 = _______
30. 8.3 + .7 = _______

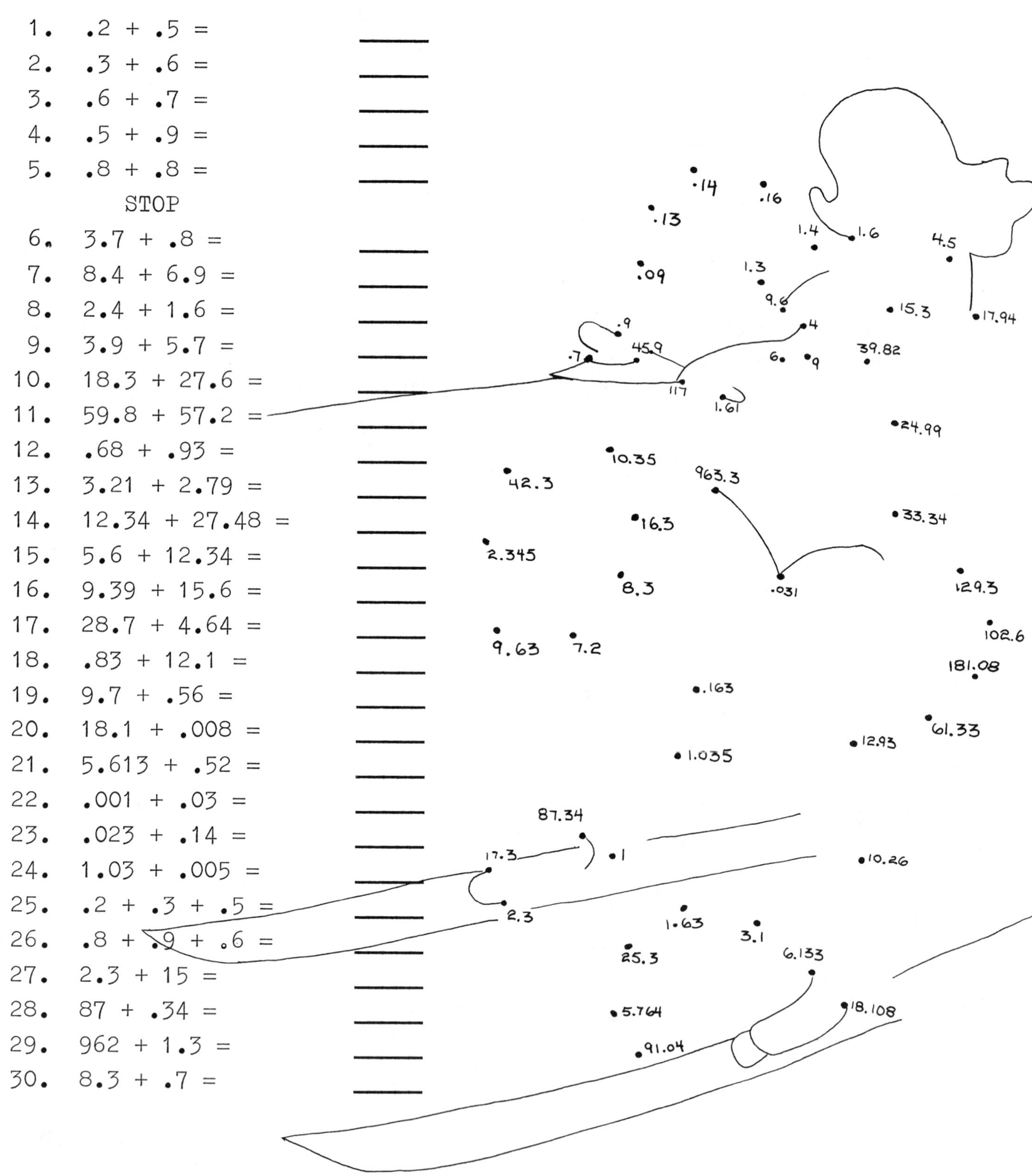

Prime Minister of India:

$$\overline{7}.\overline{5}\ \overline{8}\ \overline{7}\ \overline{3.0}\ \overline{4} \qquad .\overline{3}\ \overline{4}.\overline{5}\ \overline{8}\ \overline{6}\ \overline{7}$$

Former Prime Minister of Israel:

$$.\overline{3}\ \overline{4}.\overline{1}\ \overline{8}\ \overline{4} \qquad \overline{4}.\overline{9}\ .\overline{2}\ \overline{7}\ \overline{3.0}$$

Solve the problems. Find the number that each letter represents.
Fill in the blanks with the letters each number represents.

1. 2.4 + 3.7 = $\underset{\overline{H}}{}$ _

2. 46.3 + 33.9 = _ _ $\underset{\overline{E}}{}$

3. 84.6 + 9.5 = _ $\underset{\overline{A}}{}$ _

4. .33 + .26 = $\underset{\overline{N}}{}$ _

5. 4.52 + 5.58 = _ _ $\underset{\overline{L}}{}$

6. 24.39 + 8.7 = _ $\underset{\llcorner R \lrcorner}{}$ _

7. 456.8 + 34.67 = _ _ _ $\underset{\overline{O}}{}$ _

8. .8 + 6.3 = $\underset{\overline{I}}{}$ _

9. 7.64 + .874 = $\underset{\overline{D}}{}$ _ _ _

10. 15 + 2.3 = _ _ $\underset{\overline{G}}{}$

11. 14.6 + 9.54 + .8 = _ $\underset{\llcorner M \lrcorner}{}$ _

SUBTRACTION OF DECIMALS

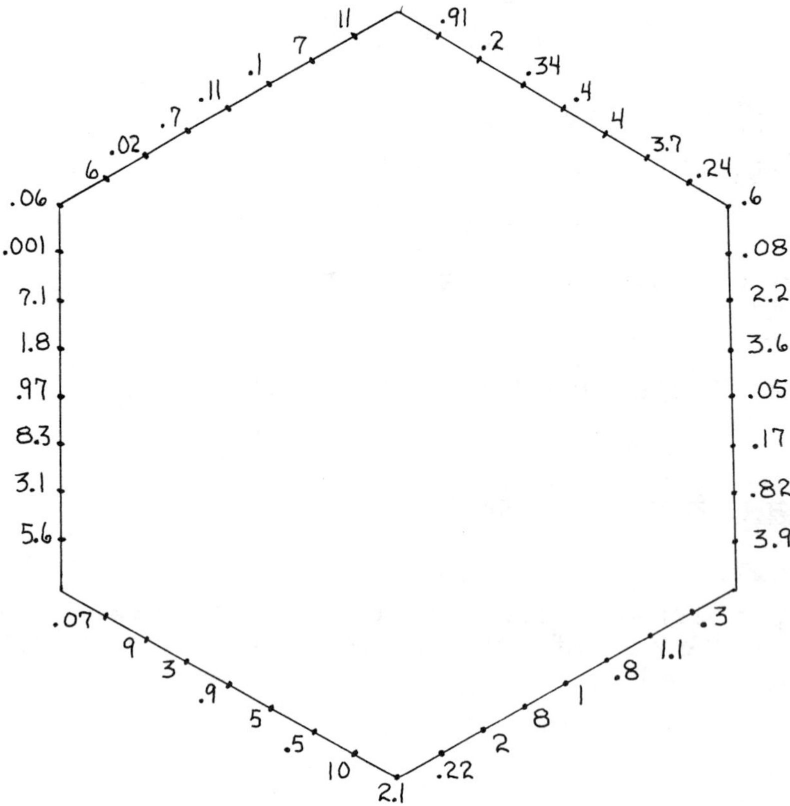

Connect "a" and "b" answers for each of the following problems:

	(a)		(b)	
1.	$.9 - .2 =$	____	$.8 - .4 =$	____
2.	$.7 - .1 =$	____	$.6 - .3 =$	____
3.	$1.2 - .3 =$	____	$2.4 - .6 =$	____
4.	$3.7 - 1.6 =$	____	$5.6 - 1.7 =$	____
5.	$.56 - .32 =$	____	$14.3 - 7.3 =$	____
6.	$2.3 - 1.3 =$	____	$18.9 - 15.3 =$	____
7.	$.46 - .26 =$	____	$16.9 - 10.9 =$	____
8.	$5.13 - 2.03 =$	____	$19.3 - 9.3 =$	____
9.	$.09 - .03 =$	____	$.14 - .07 =$	____
10.	$1.13 - .96 =$	____	$2.41 - .41 =$	____
11.	$16.4 - 9.3 =$	____	$43.5 - 40.5 =$	____
12.	$9.43 - 8.52 =$	____	$5.31 - 5.25 =$	____
13.	$2.3 - 1.2 =$	____	$.51 - .43 =$	____
14.	$5.815 - 1.815 =$	____	$5.54 - 5.43 =$	____
15.	$10 - 9.5 =$	____	$15 - 6.7 =$	____
16.	$43.7 - 41.5 =$	____	$29.73 - 28.93 =$	____
17.	$6.57 - 5.6 =$	____	$54.1 - 49.1 =$	____
18.	$82 - 78.3 =$	____	$5 - 4.9 =$	____
19.	$.87 - .869 =$	____	$44.7 - 35.7 =$	____
20.	$.9 - .08 =$	____	$4.3 - 4.08 =$	____
21.	$44.2 - 42.1 =$	____	$87 - 81.4 =$	____
22.	$99.4 - 91.4 =$	____	$.153 - .103 =$	____
23.	$.234 - .214 =$	____	$1.69 - 1.35 =$	____
24.	$36.54 - 25.54 =$	____	$2.49 - 1.89 =$	____

40

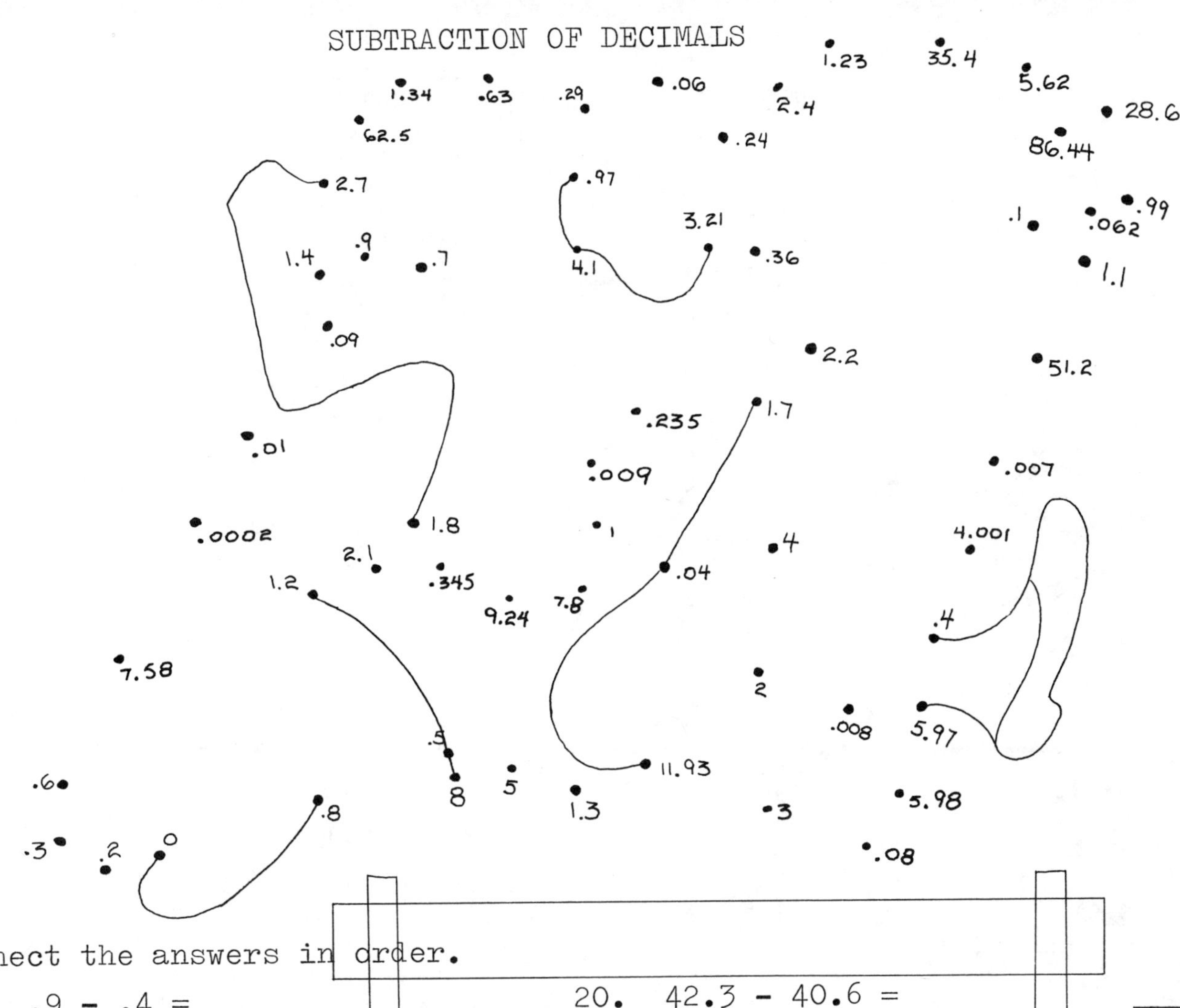

Connect the answers in order.

1.	.9 − .4 =	20.	42.3 − 40.6 =
2.	.9 − .1 =	21.	53.8 − 51.6 =
3.	.6 − .6 =	22.	291.34 − 290.24 =
4.	.8 − .6 =	23.	4.6 − 3.61 =
5.	.4 − .1 =	24.	29.3 − 29.238
6.	.9 − .3 =	25.	31 − 2.4 =
7.	2.3 − 1.1 =	26.	87 − .56 =
8.	4.4 − 2.3 =	27.	92 − 91.9 =
9.	15.7 − 13.9 =	28.	4.36 − 4 =
10.	2.13 − 2.09 =	29.	72.21 − 69 =
11.	43.7 − 39.7 =	30.	9.14 − 8.9 =
12.	87.2 − 86.8 =	31.	56.29 − 56.23 =
13.	9.14 − 3.17 =	32.	4.283 − 3.993 =
14.	2.432 − 2.424 =	33.	12.6 − 11.63 =
15.	186.3 − 184.3 =	34.	194.437 − 190.337 =
16.	21.43 − 9.5 =	35.	.843 − .143 =
17.	6.77 − 5.47 =	36.	.9 − .81 =
18.	9.6 − 4.6 =	37.	.9 − 0 =
19.	10.3 − 2.3 =	38.	26.3 − 24.9 =
	STOP	39.	8.81 − 6.11 =

MULTIPLICATION OF DECIMALS

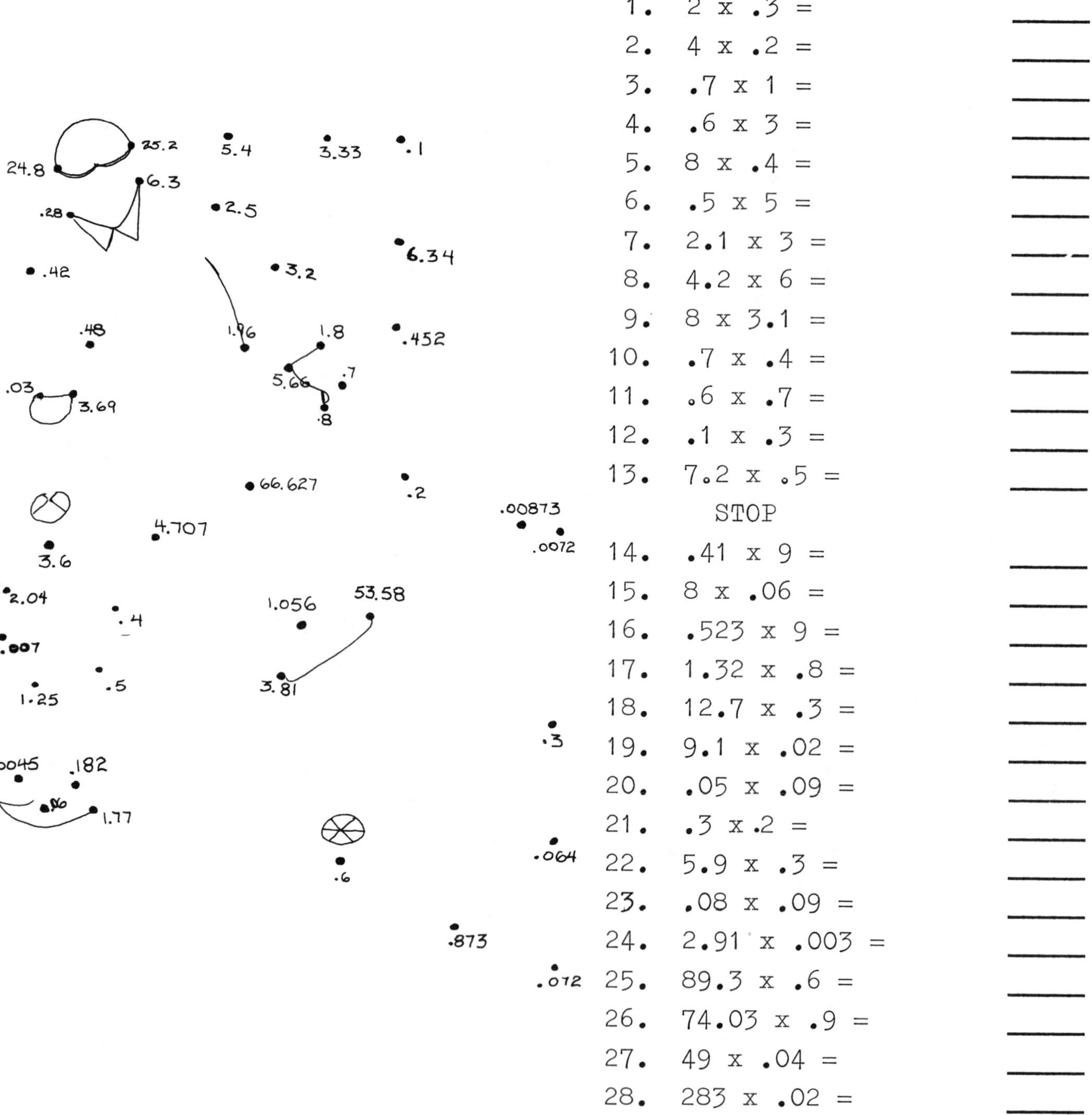

Connect the answers in order.

1. 2 x .3 = ________

2. 4 x .2 = ________

3. .7 x 1 = ________

4. .6 x 3 = ________

5. 8 x .4 = ________

6. .5 x 5 = ________

7. 2.1 x 3 = ________

8. 4.2 x 6 = ________

9. 8 x 3.1 = ________

10. .7 x .4 = ________

11. .6 x .7 = ________

12. .1 x .3 = ________

13. 7.2 x .5 = ________

STOP

14. .41 x 9 = ________

15. 8 x .06 = ________

16. .523 x 9 = ________

17. 1.32 x .8 = ________

18. 12.7 x .3 = ________

19. 9.1 x .02 = ________

20. .05 x .09 = ________

21. .3 x .2 = ________

22. 5.9 x .3 = ________

23. .08 x .09 = ________

24. 2.91 x .003 = ________

25. 89.3 x .6 = ________

26. 74.03 x .9 = ________

27. 49 x .04 = ________

28. 283 x .02 = ________

MULTIPLICATION OF DECIMALS

$\overline{1.0}$ $\overline{11}$ $\overline{2}$ $\overline{.04}$ $\overline{5}$ $\overline{.8}$ $\overline{4}$ $\overline{7}$ $\overline{.8}$ $\overline{.01}$ $\overline{11}$ $\overline{3.1}$ $\overline{.3}$ $\overline{.04}$ $\overline{.12}$ $\overline{.13}$ $\overline{4}$ $\overline{11}$ $\overline{7}$ $\overline{7}$?

Solve the problems. Find the number that each letter represents.
Fill in the blanks with the letters each number represents.

1.   .3 x 1 = $\underset{O}{\underline{}}$

2.   .7 x 4 = _ $\underset{A}{\underline{}}$

3.   .6 x .2 = $\underset{N}{[]}$

4.   2.8 x 4 = $\underset{E}{[]}$ _

5.   .263 x .5 = $\underset{T}{[]}$ _ _

6.   .021 x .9 = $\underset{R}{[]}$ _ _

7.   2.1 x 3.6 = $\underset{S}{\underline{}}$ _ _

8.   .74 x 3.2 = $\underset{C}{\underline{}}$ _ _ _

9.   8.71 x .61 = $\underset{M}{\underline{}}$ _ _ _ _

10.   .215 x 22 = $\underset{L}{\underline{}}$ _ _

11.   .146 x .32 = $\underset{I}{[]}$ _ _ _

12.   216.3 x .005 = $\underset{D}{[]}$ _ _ _

13.   48.36 x 1.1 = _ $\underset{P}{[]}$ _ _

DIVISION OF DECIMALS

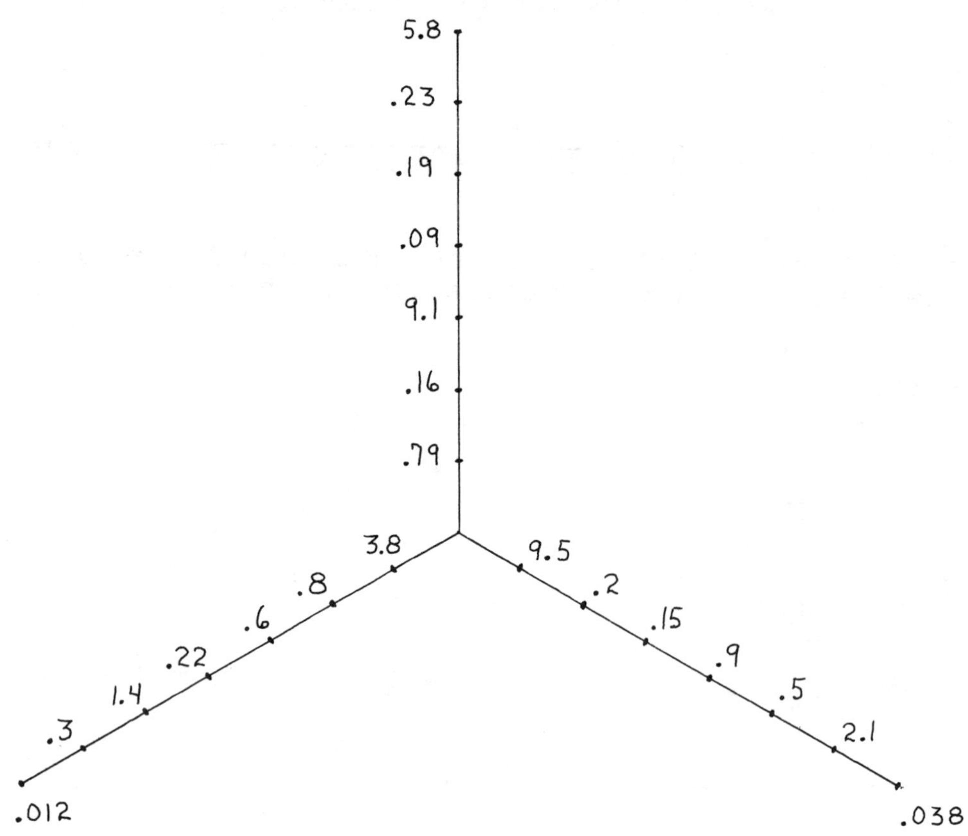

Connect "a" and "b" answers for each of the following problems:

(a)   (b)

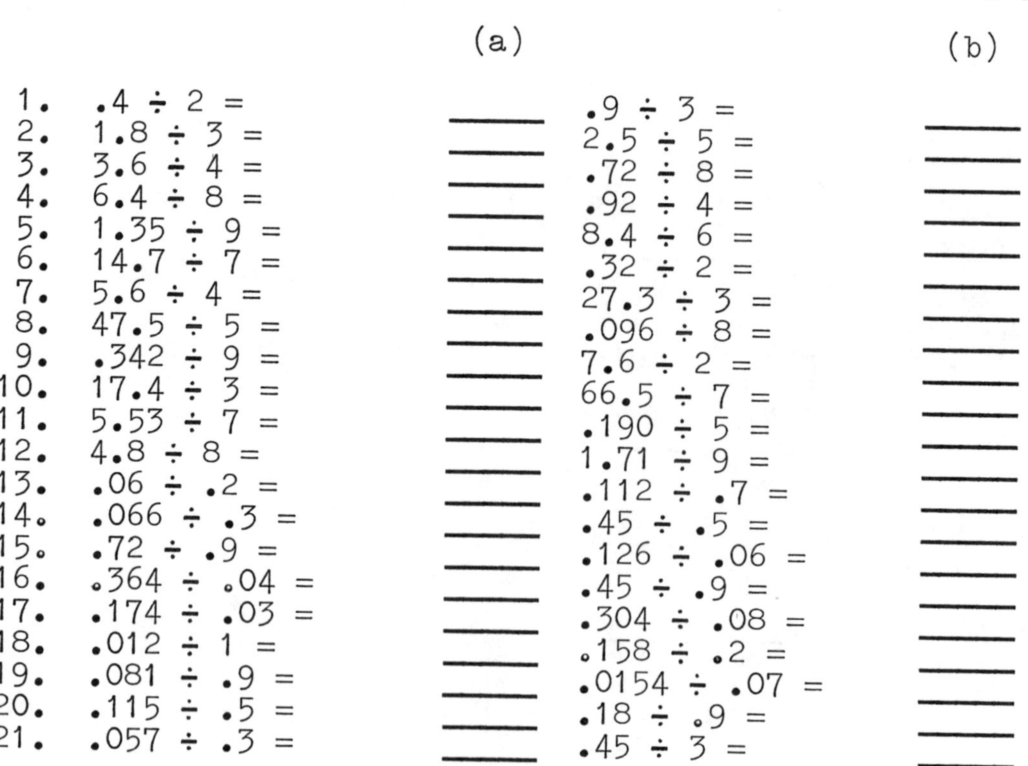

1.	.4 ÷ 2 =	.9 ÷ 3 =
2.	1.8 ÷ 3 =	2.5 ÷ 5 =
3.	3.6 ÷ 4 =	.72 ÷ 8 =
4.	6.4 ÷ 8 =	.92 ÷ 4 =
5.	1.35 ÷ 9 =	8.4 ÷ 6 =
6.	14.7 ÷ 7 =	.32 ÷ 2 =
7.	5.6 ÷ 4 =	27.3 ÷ 3 =
8.	47.5 ÷ 5 =	.096 ÷ 8 =
9.	.342 ÷ 9 =	7.6 ÷ 2 =
10.	17.4 ÷ 3 =	66.5 ÷ 7 =
11.	5.53 ÷ 7 =	.190 ÷ 5 =
12.	4.8 ÷ 8 =	1.71 ÷ 9 =
13.	.06 ÷ .2 =	.112 ÷ .7 =
14.	.066 ÷ .3 =	.45 ÷ .5 =
15.	.72 ÷ .9 =	.126 ÷ .06 =
16.	.364 ÷ .04 =	.45 ÷ .9 =
17.	.174 ÷ .03 =	.304 ÷ .08 =
18.	.012 ÷ 1 =	.158 ÷ .2 =
19.	.081 ÷ .9 =	.0154 ÷ .07 =
20.	.115 ÷ .5 =	.18 ÷ .9 =
21.	.057 ÷ .3 =	.45 ÷ 3 =

DIVISION OF DECIMALS

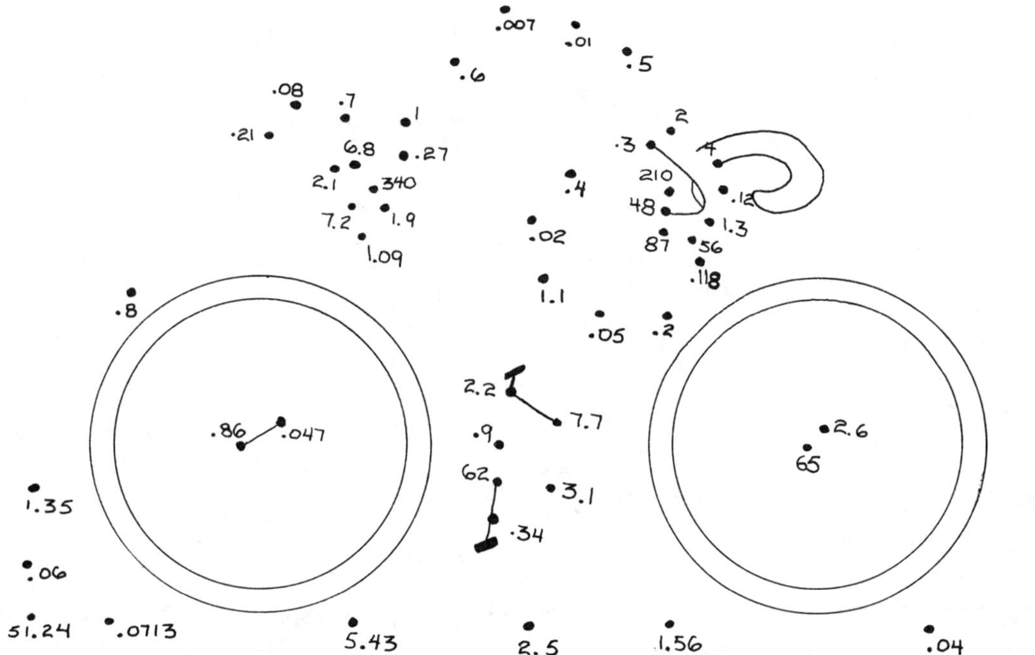

Connect the answers in order.

1. $4.2 \div 6 =$	____	
2. $.72 \div 9 =$	____	
3. $.84 \div 4 =$	____	
4. $16.8 \div 8 =$	____	
5. $14.4 \div 2 =$	____	
6. $6.02 \div 7 =$	____	
7. $.34 \div 1 =$	____	
8. $15.5 \div 5 =$	____	
9. $23.1 \div 3 =$	____	
10. $43.4 \div .7 =$	____	
11. $.218 \div .2 =$	____	
12. $.0141 \div .3 =$	____	
13. $.132 \div .06 =$	____	
14. $.045 \div .05 =$	____	
15. $1.52 \div .8 =$	____	
16. $4.32 \div .09 =$	____	
17. $34.8 \div .4 =$	____	

18. $5.6 \div .1 =$	____	
19. $1.8 \div 2 =$	____	
20. $.62 \div .01 =$	____	
21. $.708 \div 6 =$	____	
22. $4.55 \div .07 =$	____	
23. $.026 \div .01 =$	____	
24. $13.0 \div 10 =$	____	
25. $1.44 \div 12 =$	____	
26. $10.0 \div 2.5 =$	____	
27. $.30 \div .15 =$	____	
28. $.0015 \div .005 =$	____	
29. $4.2 \div .02 =$	____	
30. $.34 \div .001 =$	____	
31. $.136 \div .02 =$	____	
32. $1.35 \div 5 =$	____	
33. $5.5 \div 5.5 =$	____	
34. $.7 \div 1 =$	____	

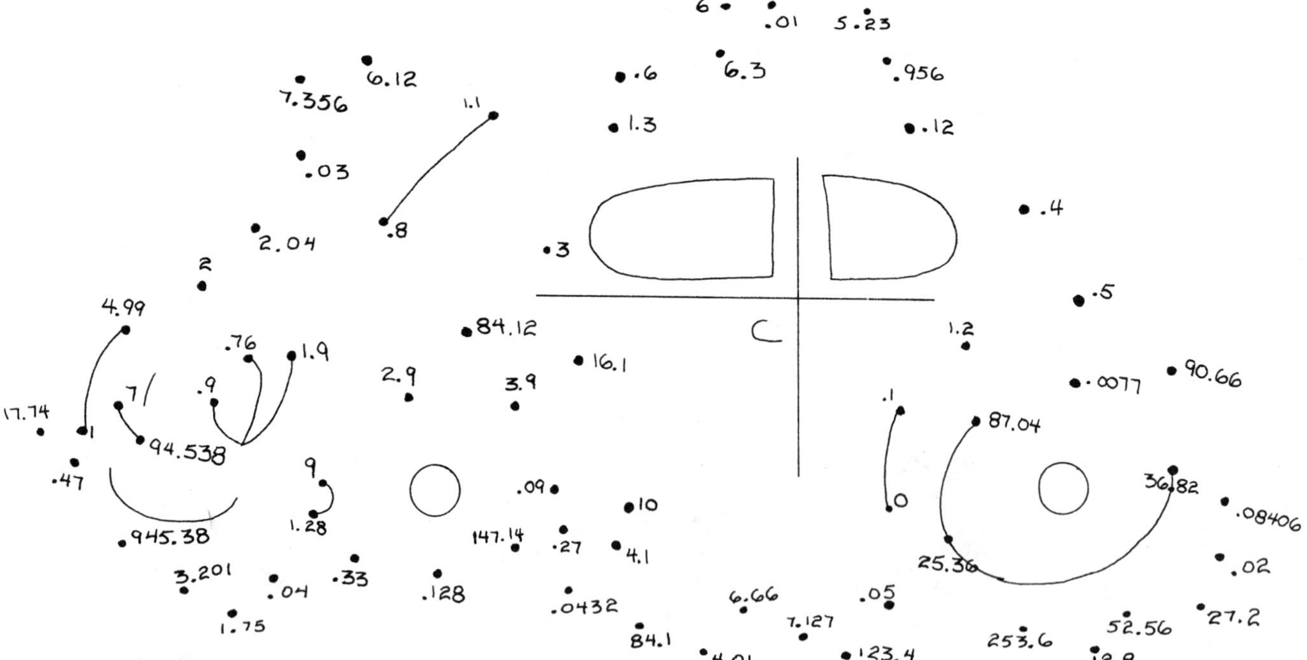

Connect the answers in order.

1.	.4 + .1 =	23.	2.375 x 0 =
2.	.7 − .3 =	24.	12.5 x .8 =
3.	.6 x .2 =	25.	96.6 ÷ 6 =
4.	1.8 ÷ 3 =	26.	82.1 + 1.3 + .72 =
5.	.2 + .3 + .6 =	27.	.38 ÷ .2 =
6.	2.4 − 1.1 =	28.	693.1 − 692.34 =
7.	.2 x 15 =	29.	.225 x 4 =
8.	1.6 ÷ 2 =	30.	94.1 + .438 =
9.	27.6 − 25.6 =	31.	17.3 − 8.3 =
10.	1.2 + 3.79 =	32.	1.45 ÷ .5 =
11.	3.5 + 3.5 =	33.	2.6 x 1.5 =
12.	.9 ÷ .9 =	34.	.348 − .258 =
13.	2.34 + 15.4 =		STOP
14.	2.35 ÷ 5 =	35.	89.4 − 2.36 =
15.	.32 x 4 =	36.	.55 x .014 =
16.	.563 − .233 =	37.	46.3 − 9.48 =
17.	.8 x .16 =	38.	9.34 x .009 =
18.	146.8 + .34 =	39.	91 − .34 =
19.	.3 x .3 =	40.	5 x .1 =
20.	.81 ÷ 3 =	41.	.9 + .1 + .2 =
21.	23.7 − 19.6 =	42.	.8 ÷ 8 =
22.	24 + 1.36 =		

DECIMAL OPERATIONS

(The decimal point should be in the box with the tenths digit.)

Across

1. 93.7 + 22.6
5. 115.9 x .5
9. 56.3 − 29.1
10. 1.6 ÷ 8
11. .236 ÷ 2
12. 1.9 x .3
13. .341 + .452
15. 2.63 − 1.53
16. 2.7 ÷ .9
17. .56 ÷ 4
18. 4.2 + .8 + .7
20. .36 ÷ .06
21. .7 x .5
22. 4.8 ÷ 6
23. 7.2 ÷ .6
25. 4.3 + 3.7
26. 86.7 − 79.3
28. 2.8 x .3
29. 56.9 − 47.9
30. 29.4 − 28.56
32. 12.9 x .6
34. .96 ÷ .3
35. 4.1 x 1.8
37. .81 ÷ .09
38. 4.6 + .73
39. 92.6 + 4.38
41. 2.3 x .57

Down

1. 32.3 − 19.77
2. 35.4 ÷ .2
3. 24.8 ÷ 4
4. 6.09 − 5.79
5. 1.6 + 3.4
6. 4.39 + 2.71
7. 1.233 − .322
8. 7.27 x 8
10. .87 ÷ 3
13. 3.7 x .2
14. 10.5 ÷ .3
17. 4.3 − 4.143
19. .23 + .484
21. .36 ÷ 1.2
22. 4.03 − 3.23
24. 1.4 + .35 + .25
25. 8.47 + .26 + .149
27. .047 ÷ .1
28. .21 x 4
29. 3.077 x 3
31. 3.15 + 1.21
33. 2.39 − 1.6
34. 13.24 ÷ 4
36. 8 + .9
38. .53 ÷ .1
40. 2.7 + 3.2 + 2.1
41. 2.34 ÷ 2.34

Write the following numbers using Roman Numerals.

Across

1. 12
4. 100
5. 54
8. 9
9. 151
11. 6
12. 10
13. 619
15. 1
16. 200
17. 5
18. 20
20. 35
22. 1·016
24. 20
25. 1000
26. 6
27. 1000
28. 36
31. 500
32. 600
34. 201
35. 1100
36. 109
38. 5
39. 590

Down

1. 19
2. 9
3. 1
4. 165
5. 50
6. 4
7. 7
9. 200
10. 2
13. 630
14. 36
16. 120
19. 16
20. 10
21. 5
22. 1000
23. 1
25. 1095
27. 1600
29. 90
30. 6
31. 700
33. 101
35. 1010
37. 10
39. 500

ROMAN NUMERALS

Cut out the squares. Fit them together so that the touching
edges name the same number.

<table>
<tr><td>26</td><td>1090</td><td>1210</td><td>612</td><td>IV</td></tr>
<tr><td>XC   40</td><td>1001   60</td><td>506   93</td><td>59   110</td><td>XVII   105</td></tr>
<tr><td>19</td><td>DCXII</td><td>21</td><td>MCX</td><td>MXC</td></tr>
<tr><td>24</td><td>XIX</td><td>52</td><td>1110</td><td>XLIX</td></tr>
<tr><td>CX   DIX</td><td>56   CVI</td><td>LVII   74</td><td>XL   CC</td><td>L   90</td></tr>
<tr><td>69</td><td>76</td><td>CDXII</td><td>CCXC</td><td>CMXC</td></tr>
<tr><td>M</td><td>160</td><td>6</td><td>XXI</td><td>412</td></tr>
<tr><td>609   87</td><td>DXXVI   MM</td><td>554   LIX</td><td>2000   8</td><td>DIII   1600</td></tr>
<tr><td>MMDIX</td><td>83</td><td>XXVI</td><td>MCCC</td><td>MMLIV</td></tr>
<tr><td>LXIX</td><td>47</td><td>990</td><td>2509</td><td>290</td></tr>
<tr><td>200   DVI</td><td>LXXIV   17</td><td>73   LVI</td><td>MCD   103</td><td>106   526</td></tr>
<tr><td>CLX</td><td>CIV</td><td>3</td><td>CDX</td><td>LXXX</td></tr>
<tr><td>104</td><td>310</td><td>410</td><td>MLV</td><td>2054</td></tr>
<tr><td>MDC   MI</td><td>LX   1400</td><td>509   CIII</td><td>CV   DCIX</td><td>33   DLIV</td></tr>
<tr><td>VI</td><td>XXIV</td><td>MCCX</td><td>CCCX</td><td>49</td></tr>
</table>

What will the computer display when you run this program?

```
 10 PRINT "WHEN STUDYING"
 20 GOTO 110
 30 PRINT "FOR A TEST"
 40 GOTO 170
 50 GOTO 90
 60 PRINT "LET THIS YOUR"
 70 PRINT "MOTTO BE"
 80 GOTO 140
 90 PRINT "TRY"
100 GOTO 40
110 GOTO 50
120 PRINT "SOME"
130 GOTO 80
140 PRINT "THINKING"
150 GOTO 200
160 PRINT "MUSIC"
170 GOTO 130
180 PRINT "AND FORGET"
190 PRINT "ABOUT TV"
200 END
```

BASIC: PRINT, LET, STRINGS

What will the computer display when you run the following program?


```
10 PRINT "KNOCK KNOCK! WHO'S THERE?"
20 LET A$ = "WHO"
30 LET B$ = "KENNETH"
40 LET C$ = "LATER"
50 LET D$ = "DOCTOR"
60 LET E$ = "LIZETTE"
70 LET F$ = "RIGHT"
80 LET G$ = "POSTPONED"
90 LET H$ = "ELSIE"
100 LET I$ = "TEST BE"
110 LET J$ = "HOUSE"
120 LET K$ = "THERE IS"
130 LET L$ = "CANDY"
140 LET M$ = "U"
150 LET N$ = "IN THE"
160 LET O$ = "ALL"
170 LET P$ = "B"
180 LET Q$ = "SARAH"
190 PRINT Q$, A$
200 PRINT Q$, D$, N$, J$
210 PRINT B$, A$
220 PRINT B$, P$, F$
230 PRINT L$, A$
240 PRINT L$, I$, G$
250 PRINT H$, A$
260 PRINT H$, M$, C$
270 PRINT E$, A$
280 PRINT E$, O$, K$
290 END
```

BASIC: PRINT, LET, STRINGS

Use this program to help you determine the five ages.

<table>
<tr><td>

```
 10 LET A = 2*D
 20 LET B$ = "BOB"
 30 LET C$ = "IS"
 40 LET D = 18
 50 LET E$ = "ELLEN"
 60 LET F$ = "IS NOT"
 70 LET G = D-3
 80 LET H$ = "HARRY"
 90 LET I$ = "A TWIN TO"
100 LET J$ = "TWICE AS OLD AS"
110 LET K$ = "FIVE YEARS OLDER THAN"
120 LET L$ = "THREE YEARS YOUNGER THAN"
130 LET M$ = "MARY"
140 LET N$ = "NICK"
150 PRINT H$; C$; I$; E$
160 PRINT N$; C$; J$; B$
170 PRINT H$; C$; G
180 PRINT N$; F$; A
190 PRINT B$; C$; L$; E$
200 PRINT E$; C$; K$; M$
210 END
```

</td><td>

How old is Bob? _____

Ellen? _____

Harry? _____

Mary? _____

Nick? _____

</td></tr>
</table>

52

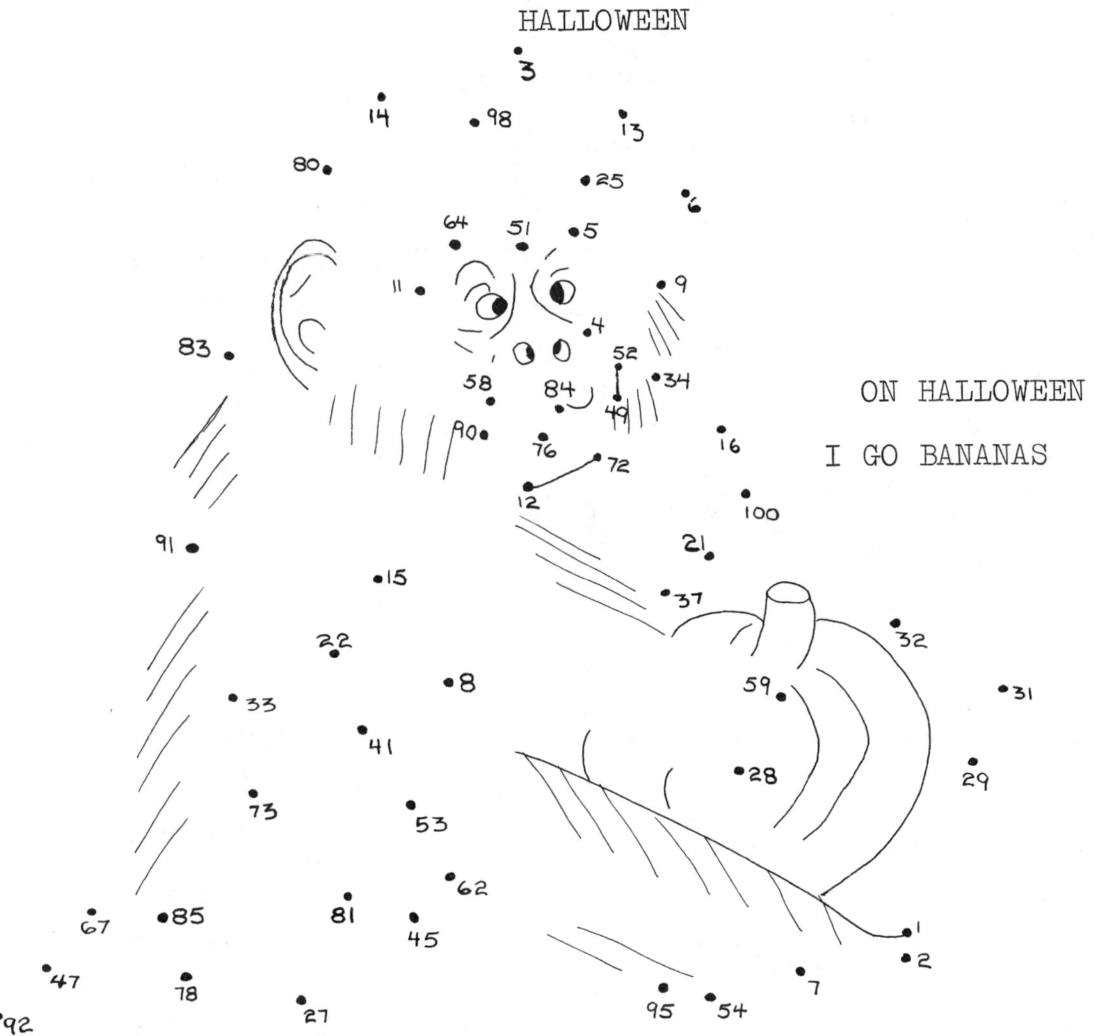

Connect the answers in order.

#	Problem	Answer
1.	5 + 7 + 3 =	_____
2.	14 − 6 =	_____
3.	7 x 4 =	_____
4.	16 + 43 =	_____
5.	29 − 17 =	_____
6.	82 + 5 + 3 =	_____
7.	53 − 42 =	_____
8.	8 x 8 =	_____
9.	36 + 8 + 7 =	_____
10.	35 ÷ 7 =	_____
11.	85 − 81 =	_____
12.	21 + 16 + 15 =	_____
13.	12 x 7 =	_____
14.	95 − 37 =	_____
15.	32 + 19 + 25 =	_____
16.	23 + 17 + 9 =	_____
17.	36 x 2 =	_____
18.	9 + 8 + 7 + 6 + 4 =	_____
19.	54 ÷ 6 =	_____
20.	42 − 17 =	_____
21.	29 + 43 + 17 + 9 =	_____
22.	9 + 28 + 7 + 36 =	_____
23.	126 − 43 =	_____
24.	287 − 196 =	_____
25.	15 + 27 + 43 =	_____
26.	13 x 6 =	_____
27.	514 − 487 =	_____
28.	12 + 19 + 28 + 36 =	_____
29.	443 − 389 =	_____
30.	63 ÷ 9 =	_____
31.	120 − 118 =	_____
32.	15 ÷ 15 =	_____

THANKSGIVING

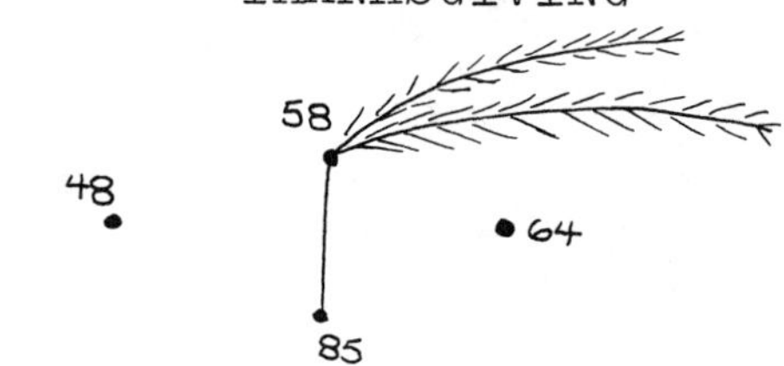

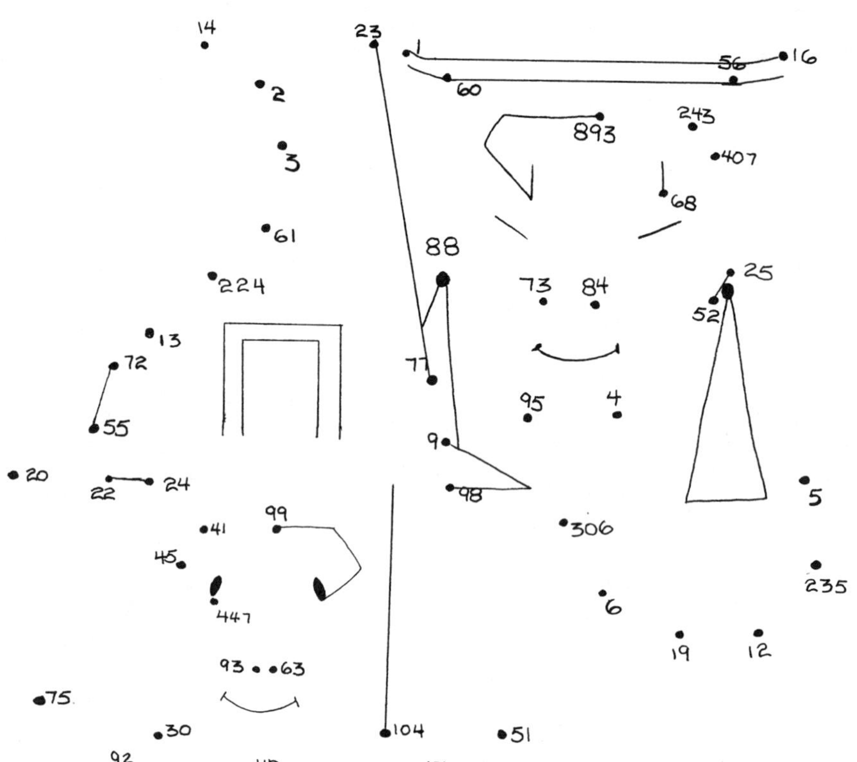

Connect the answers in order.

1. $5 + 7 + 8 + 3 =$ ___
2. $27 - 13 =$ ___
3. $9 \times 8 =$ ___
4. $18 + 37 + 22 =$ ___
5. $63 \div 7 =$ ___
6. $36 + 5 + 14 =$ ___
7. $120 \div 6 =$ ___
8. $57 - 35 =$ ___
9. $33 + 14 + 28 =$ ___
10. $23 \times 4 =$ ___
11. $120 \div 4 =$ ___
12. $17 + 29 + 38 + 26 =$ ___
13. $192 - 88 =$ ___
14. $14 + 43 + 29 + 15 =$ ___
15. $255 \div 5 =$ ___

16. $13 + 27 + 46 + 12 =$ ___
17. $73 - 49 =$ ___
18. $2 \times 3 \times 5 =$ ___
    STOP
19. $7 + 46 + 23 + 17 =$ ___
20. $256 - 193 =$ ___
21. $26 + 15 + 31 + 27 =$ ___
22. $417 - 376 =$ ___
23. $3 \times 3 \times 5 =$ ___
24. $87 + 146 + 214 =$ ___
    STOP
25. $146 \div 2 =$ ___
26. $6 \times 7 \times 2 =$ ___
27. $761 + 34 + 98 =$ ___
28. $27 \times 9 =$ ___
29. $217 + 56 + 134 =$ ___
30. $512 - 444 =$ ___
    STOP
31. $6 + 7 + 9 + 3 =$ ___
32. $43 - 27 =$ ___
33. $2 \times 4 \times 8 =$ ___
34. $23 + 16 + 19 =$ ___
35. $144 \div 3 =$ ___
36. $816 - 815 =$ ___
37. $57 + 19 + 12 =$ ___
38. $19 \times 5 =$ ___
39. $234 - 230 =$ ___
40. $16 + 23 + 13 =$ ___
41. $56 \div 1 =$ ___
42. $184 - 99 =$ ___
43. $14 + 29 + 17 =$ ___
44. $880 \div 10 =$ ___

Connect the answers in order.

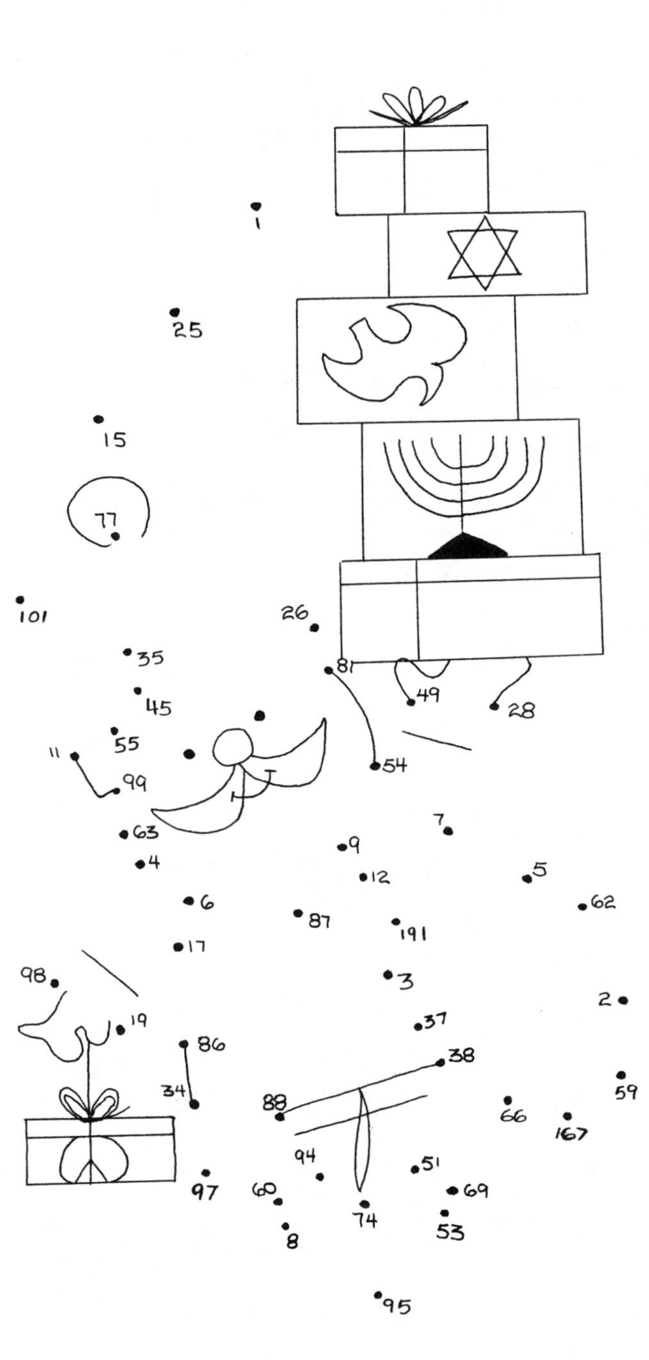

1. 11 + 7 + 8 = ________
2. 27 − 16 = ________
3. 27 + 38 + 12 = ________
4. 83 − 57 = ________
5. 27 x 3 = ________
6. 57 + 29 + 13 = ________
7. 126 ÷ 2 = ________
8. 7 x 4 − 24 = ________
9. 48 + 32 + 18 = ________
10. 92 − 73 = ________
11. 85 ÷ 5 = ________
12. 15 + 8 + 27 + 36 = ________
13. 43 x 5 − 210 = ________
14. 214 − 152 = ________
15. 136 ÷ 4 = ________
16. 31 + 20 + 19 + 18 = ________
17. 187 − 93 = ________
18. 5 x 3 x 4 = ________
19. 8 x 1 ÷ 1 = ________
20. 87 x 2 − 100 = ________
21. 81 ÷ 3 + 26 = ________
22. 123 + 289 − 343 = ________
23. 43 x 7 − 250 = ________
24. 871 − 833 = ________
25. 434 ÷ 7 = ________
26. 876 + 573 − 1442 = ________
27. 12 x 17 − 176 = ________
28. 1436 − 1387 = ________
29. 296 ÷ 8 + 17 = ________
30. 369 + 148 − 508 = ________
31. 85 x 23 − 1943 = ________
32. 510 ÷ 6 + 2 = ________
33. 929 + 236 − 1159 = ________
34. 146 x 9 − 1310 = ________

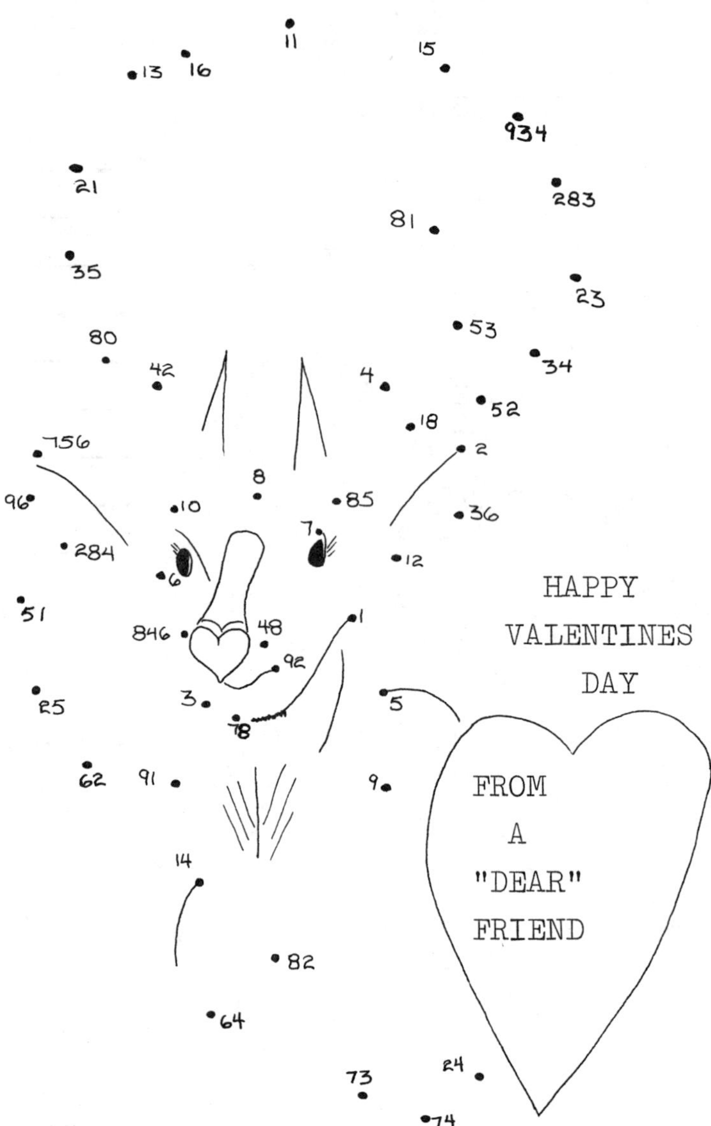

Connect the answers in order.

1. $234 + 456 - 683 =$ ___________
2. $8 \times 2 \times 3 =$ ___________
3. $276 \div 3 =$ ___________
4. $13 \times 3 \times 2 =$ ___________
5. $573 + 829 - 1399 =$ ___________
6. $47 \times 9 \times 2 =$ ___________
7. $1438 + 2769 - 4201 =$ ___________
8. $568 \div 2 =$ ___________
9. $435 \div 5 + 9 =$ ___________
10. $9 \times 7 \times 4 \times 3 =$ ___________
11. $2738 - 2143 - 585 =$ ___________
12. $264 \div 4 + 14 =$ ___________
13. $42{,}863 + 9{,}821 - 52{,}671 =$ ___________
14. $840 \div 20 =$ ___________
15. $27 \times 15 - 397 =$ ___________
16. $9624 \div 6 - 1600 =$ ___________
17. $30 \times 27 \div 10 =$ ___________
18. $12{,}436 + 21{,}879 - 34{,}300 =$ ___________
19. $5300 \div 100 =$ ___________
20. $26 \times 32 - 814 =$ ___________
21. $765 \div 9 =$ ___________
22. $19{,}436 + 8{,}215 - 27{,}649 =$ ___________
23. $16 \times 18 \div 8 =$ ___________
24. $973 - 745 - 216 =$ ___________
25. $87 \times 45 - 3914 =$ ___________
26. $833 \div 7 - 114 =$ ___________
27. $2347 + 6921 - 9259 =$ ___________
28. $272 \div 8 + 48 =$ ___________
29. $126 \times 37 - 4648 =$ ___________
30. $546 \div 2 \div 3 =$ ___________
31. $33 \div 11 =$ ___________

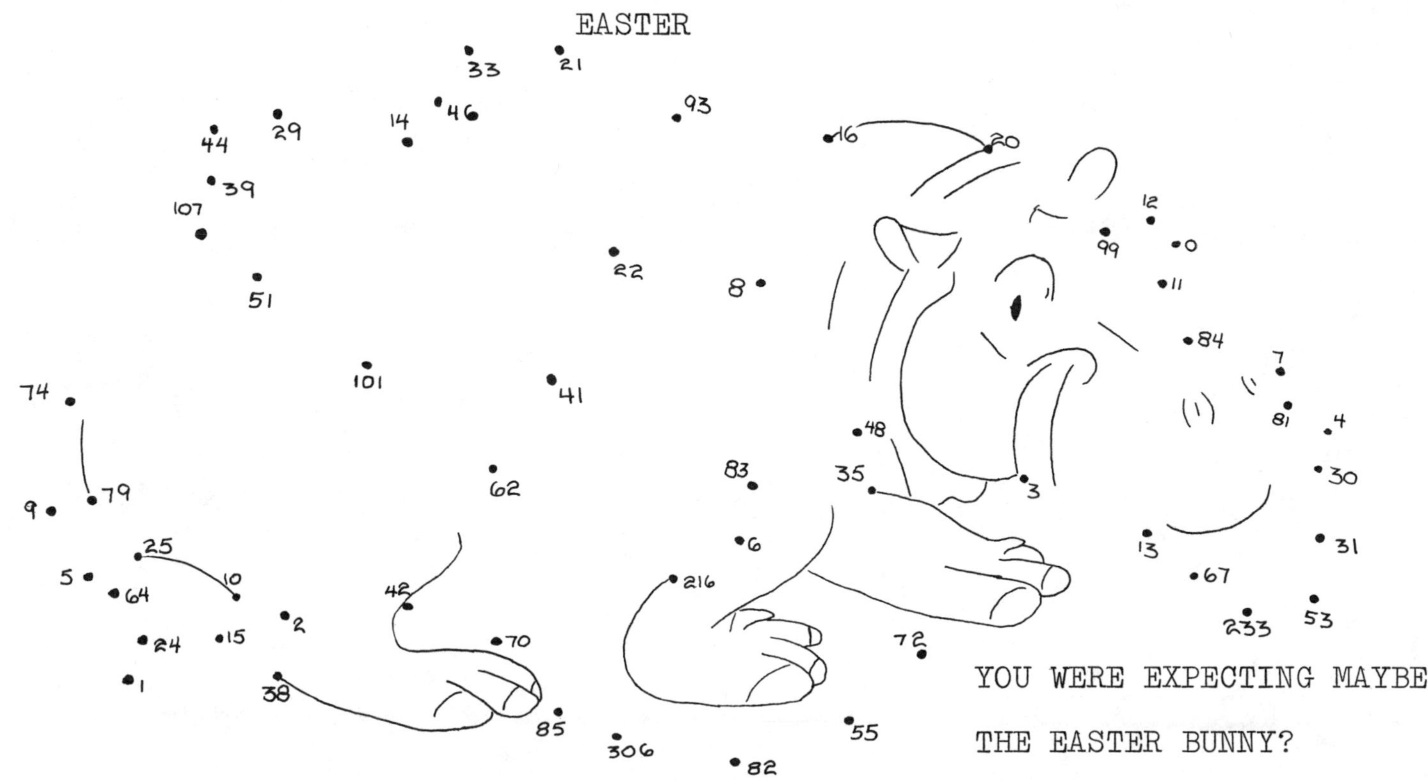

Connect the answers in order.

#	Problem		#	Problem	
1.	263 + 573 − 830 =	____	20.	23 x 46 x 2 − 2114 =	____
2.	128 − 93 =	____	21.	420 ÷ 20 x 2 =	____
3.	8 x 3 x 2 =	____	22.	931 + 139 − 1000 =	____
4.	129 + 248 − 369 =	____	23.	45 x 5 − 9 =	____
5.	12 x 18 − 200 =	____	24.	876 − 793 =	____
6.	837 ÷ 9 =	____		STOP	
7.	287 − 276 + 3 =	____	25.	33 x 11 − 360 =	____
8.	535 ÷ 5 =	____	26.	934 + 379 − 1300 =	____
9.	1273 − 1199 =	____	27.	73 x 5 + 365 − 700 =	____
10.	876 − 538 − 329 =	____	28.	476 ÷ 7 − 64 x 1 =	____
11.	27 x 15 − 400 =	____	29.	361 − 295 + 15 =	____
12.	237 ÷ 3 =	____	30.	86 x 65 − 5583 =	____
13.	614 + 382 + 126 − 1097 =	____	31.	284 + 716 − 916 =	____
14.	280 ÷ 10 + 36 =	____	32.	812 − 700 − 101 =	____
15.	963 − 774 − 165 =	____	33.	215 x 19 x 0 =	____
16.	53 x 27 − 1430 =	____	34.	515 ÷ 515 x 12 =	____
17.	1836 − 1798 =	____	35.	915 − 815 − 1 =	____
18.	512 ÷ 8 − 49 =	____	36.	436 + 564 − 980 =	____
19.	1530 + 2839 − 4359 =	____			

READING AND WRITING WHOLE NUMBERS

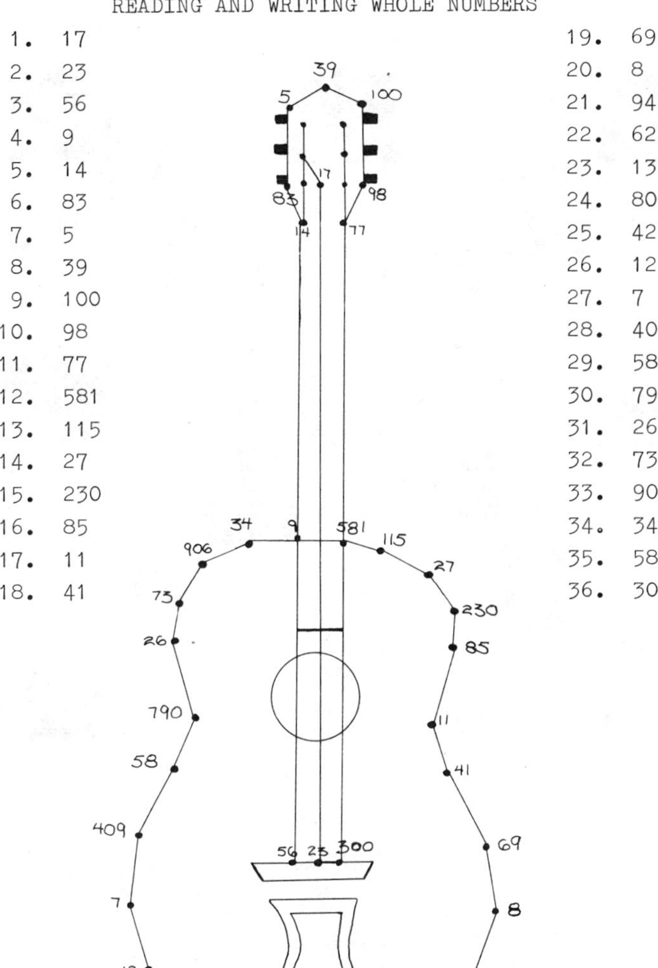

1. 17
2. 23
3. 56
4. 9
5. 14
6. 83
7. 5
8. 39
9. 100
10. 98
11. 77
12. 581
13. 115
14. 27
15. 230
16. 85
17. 11
18. 41
19. 69
20. 8
21. 94
22. 622
23. 13
24. 800
25. 42
26. 12
27. 7
28. 409
29. 58
30. 790
31. 26
32. 73
33. 906
34. 34
35. 581
36. 300

1A

READING AND WRITING WHOLE NUMBERS

1	2	3	4	5	6	7	8	9	10	11
■	(1) 1	(2) 3	(3) 5	■	(4) 7	■	(5) 6	(6) 4	(7) 3	■
(8) 9	■	(9) 8	0	(10) 3	■	(11) 4	4	0	■	(12) 8
(13) 9	(14) 3	■	(15) 4	0	(16) 3	8	0	■	(17) 2	0
(18) 9	6	(19) 9	■	(20) 5	7	3	■	(21) 8	0	0
(22) 9	4	3	(23) 1	■	(24) 2	■	(25) 2	5	1	4
■	■	(26) 1	2	(27) 4	■	(28) 9	3	7	■	■
(29) 8	(30) 3	7	6	■	(31) 3	■	(32) 5	0	(33) 4	(34) 6
(35) 7	0	6	■	(36) 5	6	(37) 2	■	(38) 9	8	0
(39) 9	7	■	(40) 5	7	3	8	(41) 7	■	(42) 4	3
(43) 6	■	(44) 8	2	1	■	(45) 6	6	(46) 2	■	(47) 6
■	(48) 3	4	3	■	(49) 2	■	(50) 8	0	(51) 9	■

2A

ADDITION OF WHOLE NUMBERS

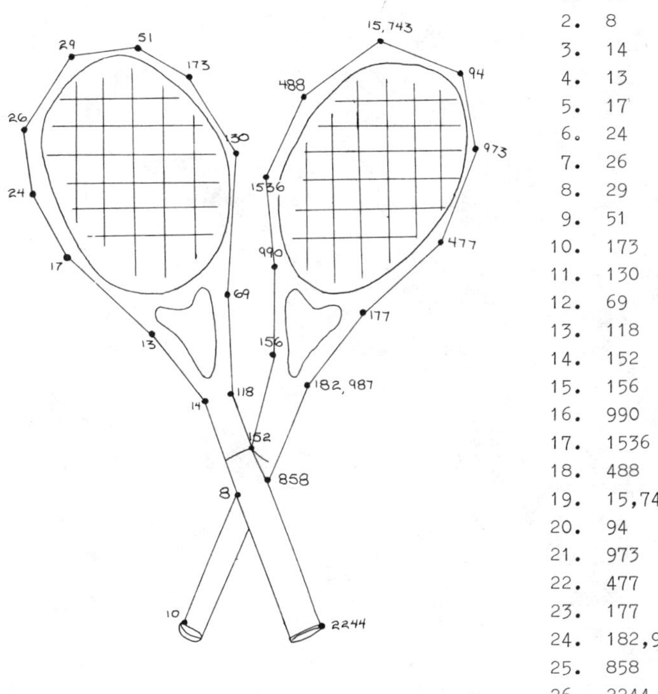

1. 10
2. 8
3. 14
4. 13
5. 17
6. 24
7. 26
8. 29
9. 51
10. 173
11. 130
12. 69
13. 118
14. 152
15. 156
16. 990
17. 1536
18. 488
19. 15,743
20. 94
21. 973
22. 477
23. 177
24. 182,987
25. 858
26. 2244

3A

ADDITION OF WHOLE NUMBERS

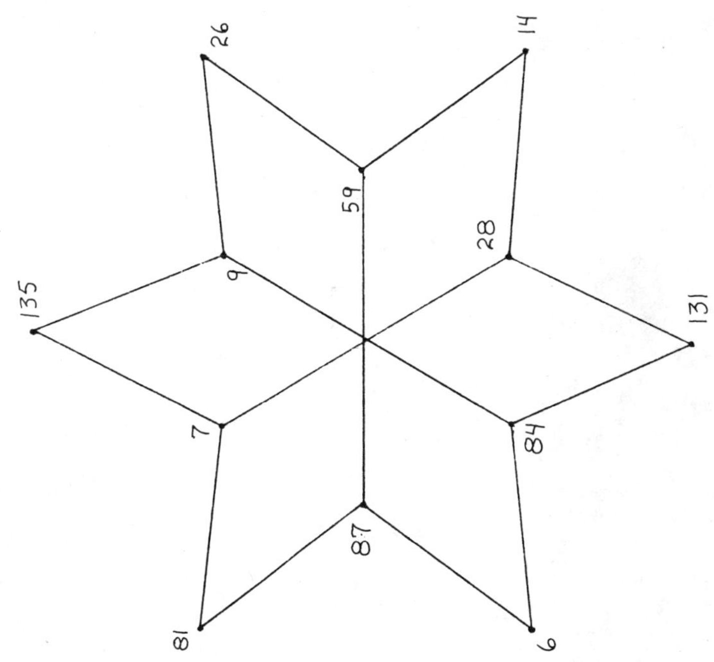

	(a)	(b)
1.	14	28
2.	9	84
3.	81	7
4.	6	87
5.	59	26
6.	87	59
7.	135	9
8.	131	28
9.	7	28
10.	131	84
11.	7	135
12.	84	6
13.	9	26
14.	81	87
15.	14	59

4A

5A

7A

ADDITION OF WHOLE NUMBERS

A D D I T I O N   G E T S   I T   T O G E T H E R !
10 7 7 8 3 8 4 1   11 5 3 9   8 3   3 4 11 5 3 6 5 2

1. 2 4 4   **244**
    R

2. 2 6 6 9   **2669**
    H

3. 4 4 9   **449**
    S

4. 3 5 1 4   **3514**
    N

5. 4 5 0 0   **4500**
    O

6. 1 3 5 5   **1355**
    E

7. 1 0 0 5 7 **10057**
    TA

8. 1 1 7 8 9 **11789**
    I

9. 1 4 5 1 1 1 **145111**
    TGT

10. 2 8 1 3 4 **28134**
    T

11. 3 7 8 2   **3782**
    D

6A

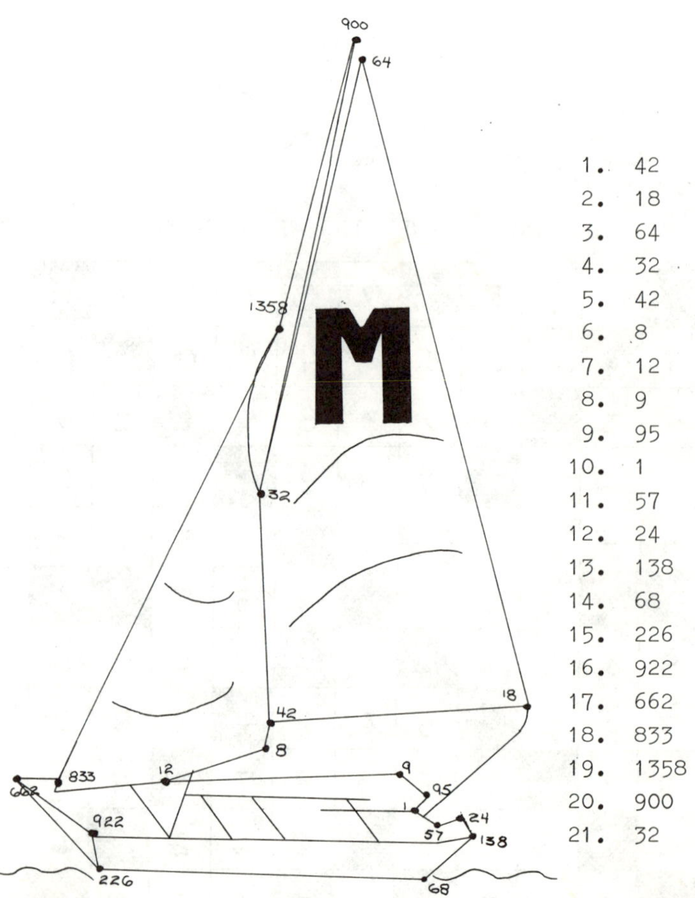

1.   42
2.   18
3.   64
4.   32
5.   42
6.   8
7.   12
8.   9
9.   95
10. 1
11. 57
12. 24
13. 138
14. 68
15. 226
16. 922
17. 662
18. 833
19. 1358
20. 900
21. 32

8A

SUBTRACTION OF WHOLE NUMBERS

The first man to set foot on the moon:

$$\frac{N}{14}\ \frac{E}{12}\ \frac{I}{4}\ \frac{L}{6} \qquad \frac{A}{11}\ \frac{R}{2}\ \frac{M}{1}\ \frac{S}{8}\ \frac{T}{7}\ \frac{R}{2}\ \frac{O}{5}\ \frac{N}{14}\ \frac{G}{3}$$

A famous Chinese teacher:

$$\frac{C}{13}\ \frac{O}{5}\ \frac{N}{14}\ \frac{F}{9}\ \frac{U}{10}\ \frac{C}{13}\ \frac{I}{4}\ \frac{U}{10}\ \frac{S}{8}$$

1. 1 2 (E)

2. 1 7 (M)

3. 7 (T)

4. 5 0 (O)

5. 8 6 3 (L)

6. 8 8 8 (S)

7. 2 4 3 (G)

8. 2 0 9 (R)

9. 8 4 (I)

10. 1 1 9 8 6 (A)

11. 1 4 9 7 8 (N)

12. 2 6 1 3 (C)

13. 6 1 0 6 (U)

14. 3 3 9 (F)

9A

SUBTRACTION OF WHOLE NUMBERS

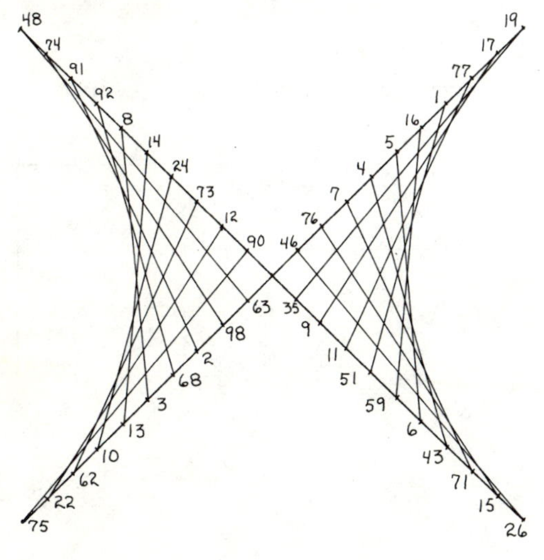

	(a)	(b)			
1.	11	77	11.	71	7
2.	14	13	12.	62	73
3.	76	15	13.	75	90
4.	92	68	14.	??	12
5.	16	59	15.	43	4
6.	48	63	16.	2	91
7.	51	1	17.	10	24
8.	19	35	18.	6	5
9.	46	26	19.	9	17
10.	8	3	20.	74	98

10A

MULTIPLICATION OF WHOLE NUMBERS

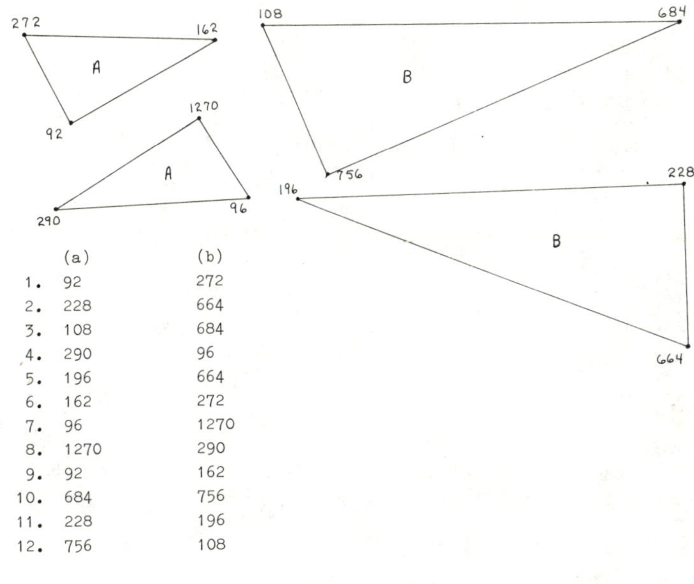

	(a)	(b)
1.	92	272
2.	228	664
3.	108	684
4.	290	96
5.	196	664
6.	162	272
7.	96	1270
8.	1270	290
9.	92	162
10.	684	756
11.	228	196
12.	756	108

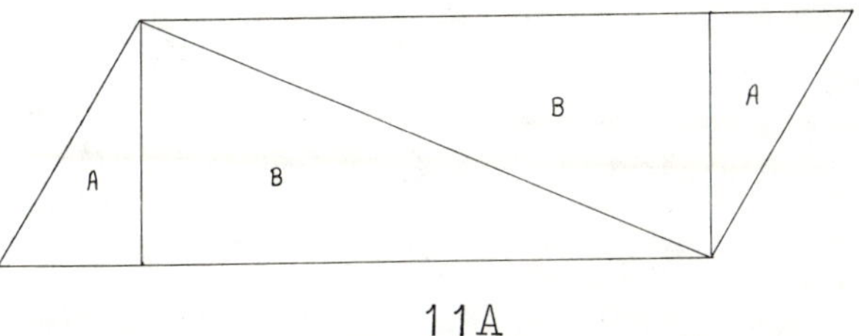

11A

MULTIPLICATION OF WHOLE NUMBERS

A famous Apache Indian, born in Arizona:

$$\frac{G}{10}\ \frac{E}{5}\ \frac{R}{8}\ \frac{O}{7}\ \frac{N}{4}\ \frac{I}{1}\ \frac{M}{9}\ \frac{O}{7}$$

A Sioux Indian Chief, born in South Dakota:

$$\frac{S}{11}\ \frac{I}{1}\ \frac{T}{2}\ \frac{T}{2}\ \frac{I}{1}\ \frac{N}{4}\ \frac{G}{10} \qquad \frac{B}{6}\ \frac{U}{3}\ \frac{L}{12}\ \frac{L}{12}$$

1. 2 2 4 (N)

2. 1 8 5 (I)

3. 5 2 2 (E)

4. 1 9 8 (R)

5. 3 2 9 (U)

6. 9 3 9 (M)

7. 3 2 5 6 (T)

8. 1 2 7 4 (L)

9. 1 1 2 0 (S)

10. 9 6 (B)

11. 1 0 7 3 6 (G)

12. 7 8 6 0 6 (O)

12A

1. 216
2. 258
3. 445
4. 504
5. 882
6. 992
7. 1311
8. 1792
9. 2820
10. 3702
11. 10284
12. 216
13. 41670
14. 0
15. 97
16. 174
17. 95
18. 7636
19. 4085
20. 82775
21. 241660
22. 184933
23. 151904
24. 778207
25. 42483
26. 280
27. 162
28. 960
29. 504

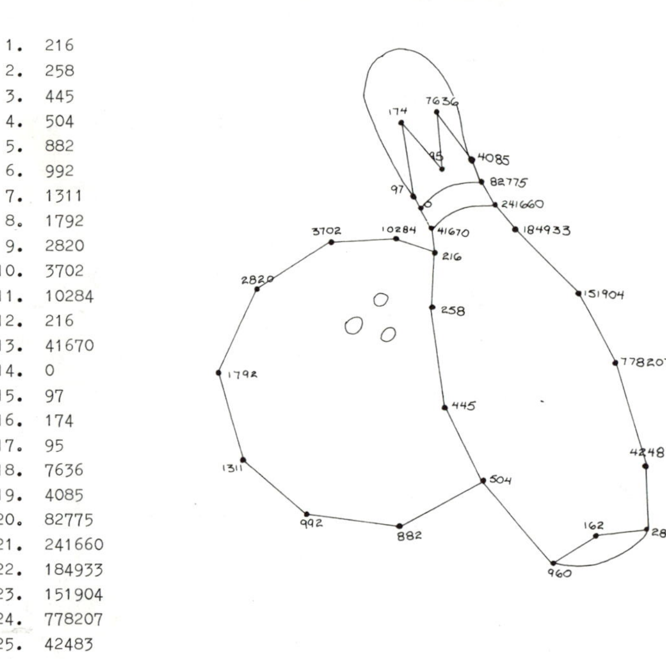

13A

MULTIPLICATION OF WHOLE NUMBERS

1	2	3	4	5	6	7	8	9	10	11
█	1	4	9	8	█	2	4	1	8	█
8	█	9	7	5	█	1	8	5	█	9
8	1	█	2	4	9	6	0	█	9	8
8	6	1	█	7	8	2	█	4	4	2
8	2	7	4	█	6	█	2	4	0	8
█	█	4	4	8	█	2	1	0	█	█
3	6	5	2	█	7	█	2	5	1	2
1	8	0	█	6	2	7	█	5	6	8
9	8	█	2	4	3	5	3	█	5	6
0	█	3	7	5	█	5	8	5	█	8
█	2	8	9	8	█	7	4	8	8	█

14A

$$\frac{D}{6}\,\frac{I}{5}\,\frac{V}{8}\,\frac{I}{5}\,\frac{D}{6}\,\frac{E}{10}\qquad\frac{A}{3}\,\frac{N}{11}\,\frac{D}{6}\qquad\frac{C}{2}\,\frac{O}{9}\,\frac{N}{11}\,\frac{Q}{1}\,\frac{U}{4}\,\frac{E}{10}\,\frac{R}{7}\ !$$

1. $\frac{4}{U}\ \frac{0}{\ }$
2. $\frac{7}{R}\ \frac{0}{\ }$
3. $\frac{9}{O}$
4. $1\ \frac{0}{[E]}$
5. $1\ \frac{5}{I}$
6. $\frac{1}{[N]}\ 1$
7. $\frac{6}{\ }\ \frac{1}{Q}$
8. $8\ \frac{3}{A}$
9. $\frac{6}{D}\ 4$
10. $4\ \frac{2}{C}$
11. $5\ \frac{8}{V}$

15A

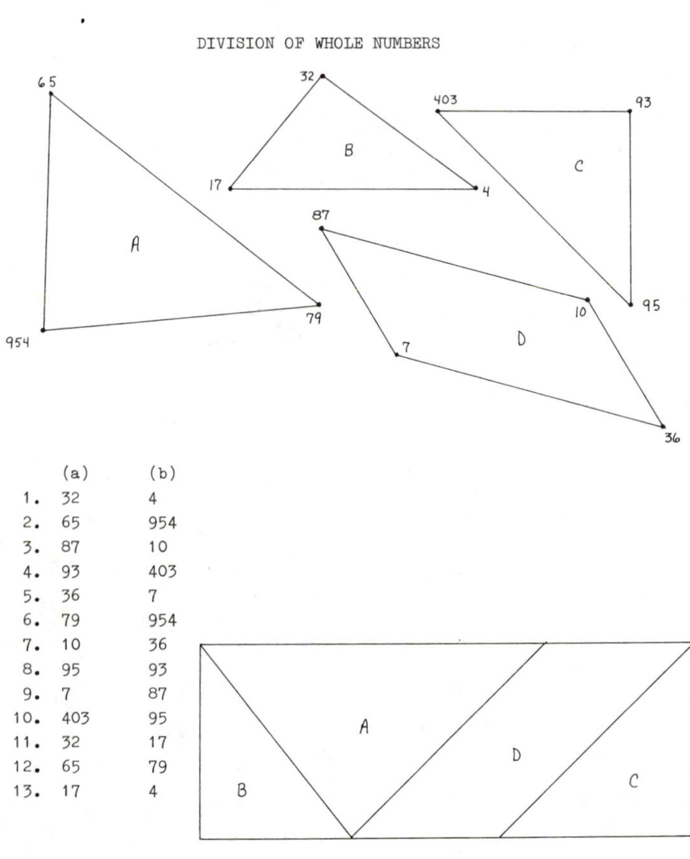

	(a)	(b)
1.	32	4
2.	65	954
3.	87	10
4.	93	403
5.	36	7
6.	79	954
7.	10	36
8.	95	93
9.	7	87
10.	403	95
11.	32	17
12.	65	79
13.	17	4

16A

DIVISION OF WHOLE NUMBERS

1:2	2:3	■	3:1	4:2	■	5:2	6:3	7:4	■	8:9
9:7	■	10:8	3	■	11:5	1	5	■	12:5	6
■	13:8	7	■	14:2	8	6	■	15:4	8	3
16:1	0	■	17:9	7	3	■	18:5	3	2	■
19:2	■	20:1	5	6	■	21:2	8	0	■	22:5
■	23:2	1	8	■	24:2	5	8	■	25:1	7
26:4	3	2	■	27:1	6	9	■	28:2	8	5
29:3	5	■	30:1	2	0	■	31:5	0	0	■
32:3	■	33:6	1	6	■	34:4	2	0	■	35:4
■	36:2	1	5	■	37:7	7	7	■	38:2	5
39:1	2	7	■	40:5	0	0	■	41:3	6	7

17A

DIVISION OF WHOLE NUMBERS

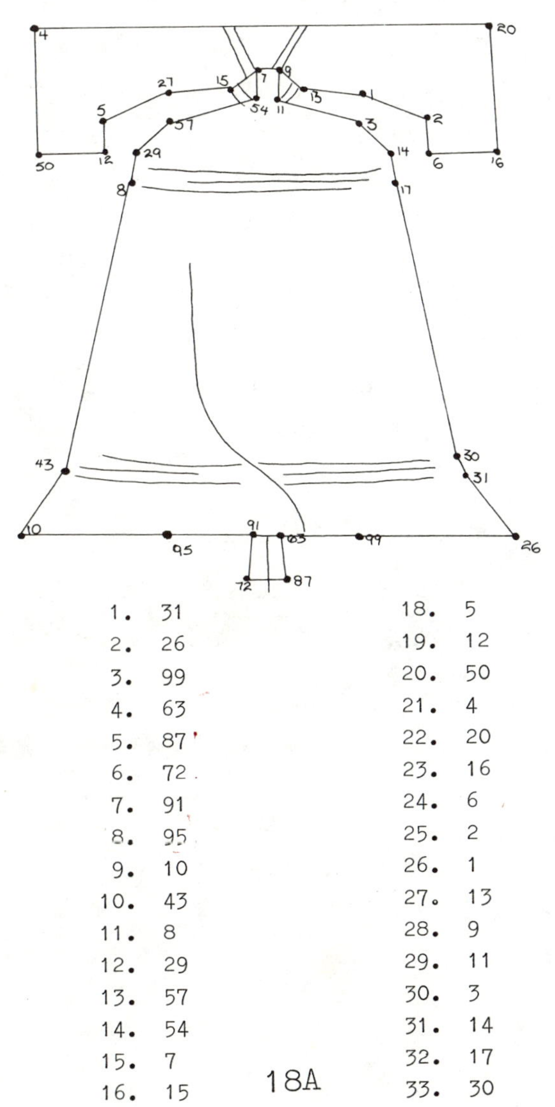

1. 31		18. 5	
2. 26		19. 12	
3. 99		20. 50	
4. 63		21. 4	
5. 87		22. 20	
6. 72		23. 16	
7. 91		24. 6	
8. 95		25. 2	
9. 10		26. 1	
10. 43		27. 13	
11. 8		28. 9	
12. 29		29. 11	
13. 57		30. 3	
14. 54		31. 14	
15. 7		32. 17	
16. 15		33. 30	
17. 27			

18A

OPERATIONS WITH WHOLE NUMBERS

Circlegram:

(2)(4)(5)   (7)(1)
(2)   (2) (1)(7)(4)(4)
(4)(1)(2) (3)(6)(8) 3 (6)
(7)   (5)   (6)   (1)   (5)
(0)(1)(2)(3)(4)(5)(6)(7)(8)(9)
(8)   (7)   (4)   (4)   (3)
(7)(5)   (3)(4)(3)   (5)
(9)(7)(4)   (2)   (5)(8)(0)
(8)

Solve all the problems below. Then fit each answer into the circlegram.

Across

$$369 + 605 = 974$$
$$892 - 647 = 245$$
$$25 \times 3 = 75$$
$$3\,)\overline{1236} = 412$$
$$436 \times 4 = 1744$$
$$128 + 235 + 217 = 580$$
$$110 + 106 + 127 = 343$$

Down

$$265 + 384 + 972 + 195 = 1816$$
$$55197 - 40539 = 14658$$
$$41147 - 16439 = 24708$$
$$6\,)\overline{2208} = 368$$
$$8262 \times 9 = 74358$$
$$14043 \times 18 = 252774$$
$$13247 - 3897 = 9350$$
$$5\,)\overline{27160} = 5432$$
$$7\,)\overline{366548} = 52364$$

19A

WHOLE NUMBER OPERATIONS

983-495	562-493	813+729	193-78	8x8
983-495 87 8x9 72÷4	562-493 72 84÷3 273+146	813+729 28 57÷19 279÷3	193-78 3 11x11 56÷4	8x8 121 57 440
18 154 13x3 12x12	419 39 200÷4 16x8	93 50 35 177	14 5x7 7 82	257+183 56÷8 93 15x8
144 6x10 47 9	128 94÷2 9x11 1992	59x3 99 72÷6 29x3	492÷6 12 10x8 17x6	120 80 4x3 595÷7
810÷90 314 86x7 210	24x83 602 500÷50 19x8	87 10 730÷10 57+92	102 73 639÷9 293-57	85 71 9x17 15x15
15x14 9x9 143 86+392	152 13x11 591÷3 27x14	149 197 138 125÷5	236 46x3 2 6x9	225 400÷200 417 426+397
478 89 297÷9 827+956	378 33 61x4 6x5	25 244 438 413-216	54 876÷2 324 817-534	823 18x18 28÷7 161+259

20A

1. 98
2. 86
3. 18
4. 97
5. 96
6. 4
7. 9
8. 6
9. 2
10. 42
11. 3
12. 11
13. 8
14. 87
15. 5
16. 1
17. 10
18. 12
19. 7
20. 73
21. 98
22. 29
23. 15
24. 91
25. 21
26. 25
27. 28
28. 69
29. 43
30. 46
31. 97

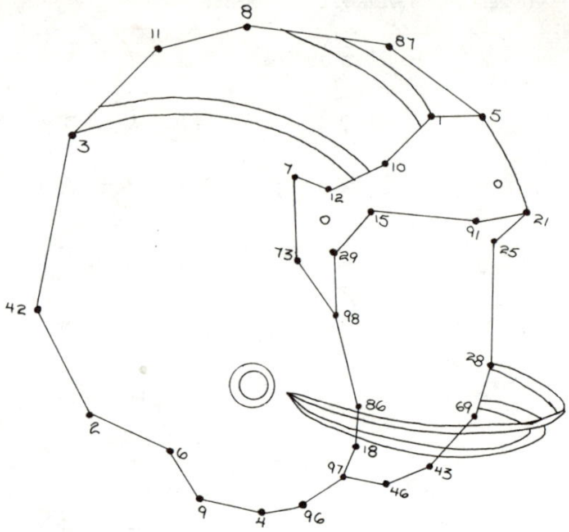

21A

EQUIVALENT FRACTIONS

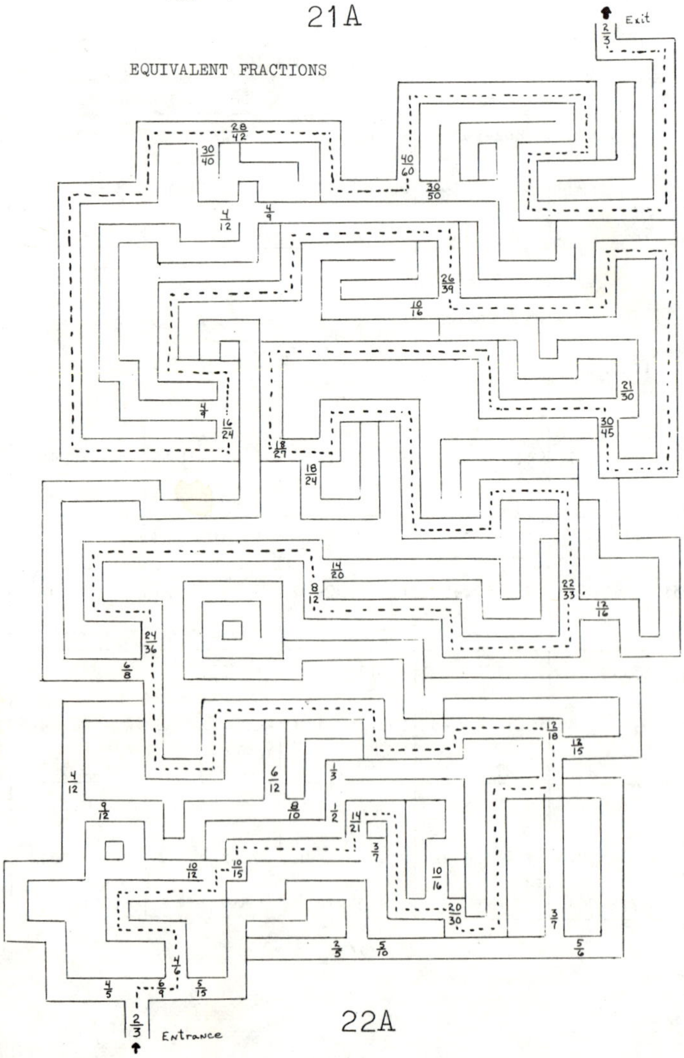

22A

ADDITION OF FRACTIONS

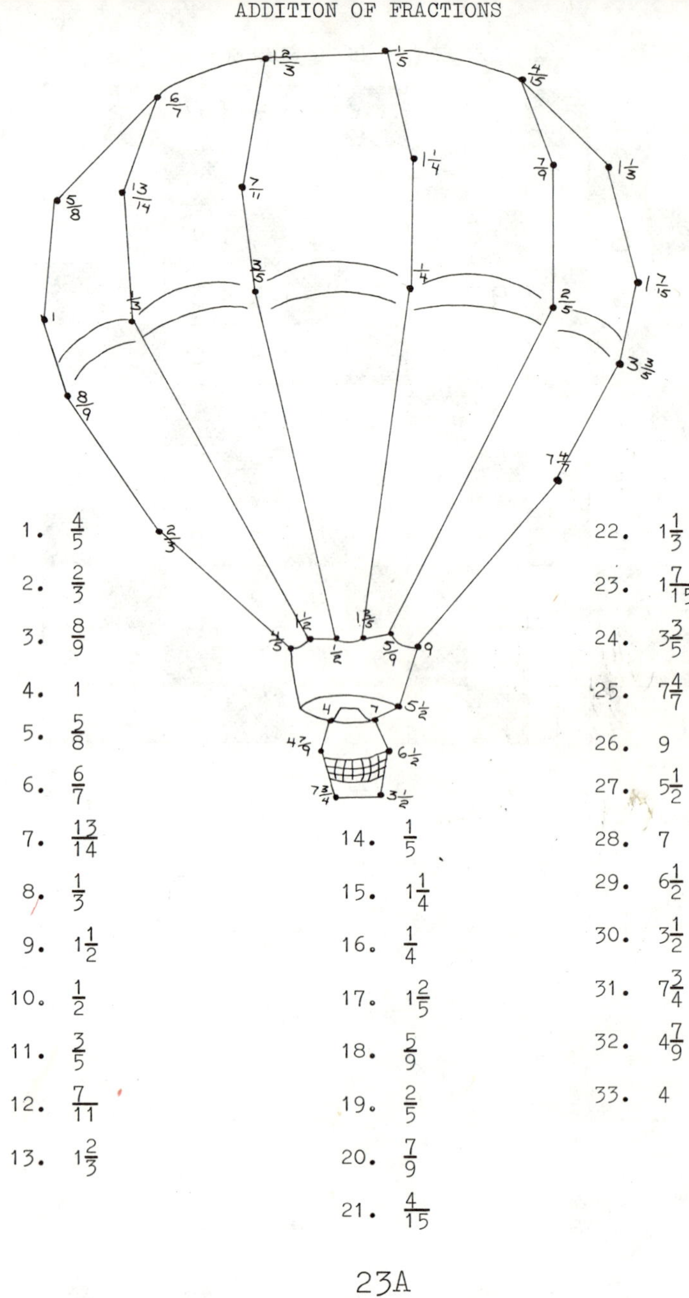

1. $\frac{4}{5}$
2. $\frac{2}{3}$
3. $\frac{8}{9}$
4. 1
5. $\frac{5}{8}$
6. $\frac{6}{7}$
7. $\frac{13}{14}$
8. $\frac{1}{3}$
9. $1\frac{1}{2}$
10. $\frac{1}{2}$
11. $\frac{3}{5}$
12. $\frac{7}{11}$
13. $1\frac{2}{3}$
14. $\frac{1}{5}$
15. $1\frac{1}{4}$
16. $\frac{1}{4}$
17. $1\frac{2}{5}$
18. $\frac{5}{9}$
19. $\frac{2}{5}$
20. $\frac{7}{9}$
21. $\frac{4}{15}$
22. $1\frac{1}{3}$
23. $1\frac{7}{15}$
24. $3\frac{3}{5}$
25. $7\frac{4}{7}$
26. 9
27. $5\frac{1}{2}$
28. 7
29. $6\frac{1}{2}$
30. $3\frac{1}{2}$
31. $7\frac{3}{4}$
32. $4\frac{7}{9}$
33. 4

23A

ADDITION OF FRACTIONS

Martin Luther King was a famous

$$\underset{13}{C}\ \underset{5}{I}\ \underset{6}{V}\ \underset{5}{I}\ \underset{23}{L} \qquad \underset{11}{R}\ \underset{5}{I}\ \underset{3}{G}\ \underset{4}{H}\ \underset{7}{T}\ \underset{8}{S} \qquad \underset{23}{L}\ \underset{12}{E}\ \underset{2}{A}\ \underset{10}{D}\ \underset{12}{E}\ \underset{11}{R}$$

1. A = 2
2. H = 4
3. S = 8
4. I = 5
5. D = 10
6. T = 7
7. E = 12
8. C = 13
9. R = 11
10. G = 3
11. L = 23
12. V = 6

24A

Other answers are possible.

Circles: (1/2) (1/3) (2/3) (1/4) (1/6) (1/5)

1. (2/3) + (1/4) + (1/2) = $1\frac{5}{12}$
2. (1/2) + (1/3) [1/4] + (1/6) [2/3] = 1
3. (1/3) + (2/3) + (1/4) = $1\frac{1}{4}$
4. (1/2) + (1/3) + (2/3) = $1\frac{1}{2}$
5. (1/3) [4] + (1/6) [4] + (1/5) = $\frac{7}{10}$
6. (1/2) [1/6 +] + (2/3) [6/3] + (1/5) = $1\frac{11}{30}$
7. (1/3) + (1/4) + (1/6) = $\frac{3}{4}$
8. (1/3) + (1/4) + (1/5) = $\frac{47}{60}$
9. (2/3) [6/3 +] + (1/4) + (1/6) [1/2] = $1\frac{1}{12}$
10. (2/3) + (1/6) + (1/5) = $1\frac{1}{30}$

25A

SUBTRACTION OF FRACTIONS

	(a)	(b)
1.	$\frac{1}{3}$	$\frac{1}{11}$
2.	$\frac{1}{7}$	$\frac{3}{5}$
3.	$\frac{1}{2}$	$\frac{2}{3}$
4.	1	$\frac{3}{5}$
5.	$\frac{1}{2}$	$\frac{1}{4}$
6.	$\frac{1}{7}$	$\frac{3}{4}$
7.	$\frac{2}{3}$	$\frac{4}{5}$
8.	$\frac{1}{4}$	$\frac{4}{5}$
9.	1	$\frac{3}{4}$

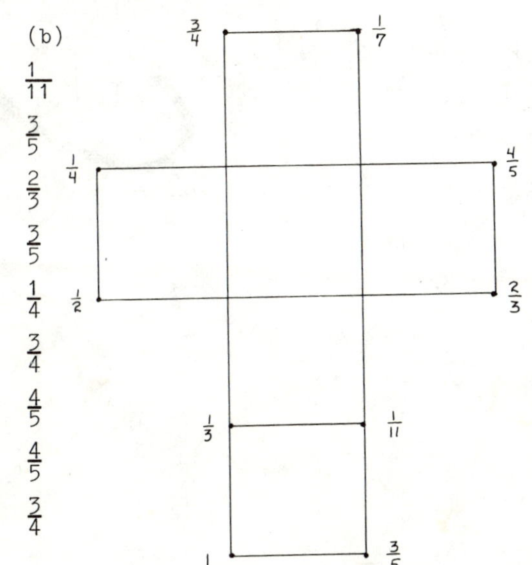

26A

1. $\frac{7}{9}$	8. $\frac{1}{12}$	23. $\frac{1}{24}$
2. $\frac{3}{8}$	9. $\frac{3}{26}$	24. $\frac{5}{24}$
3. $\frac{2}{5}$	10. $\frac{1}{8}$	25. $\frac{1}{15}$
4. $\frac{5}{7}$	11. $\frac{1}{3}$	26. $\frac{3}{14}$
5. $\frac{1}{4}$	12. $\frac{4}{9}$	27. $\frac{1}{36}$
6. $\frac{2}{7}$	13. $\frac{5}{8}$	28. $\frac{7}{16}$
7. $\frac{1}{5}$	14. $\frac{2}{3}$	29. $\frac{7}{36}$
	15. $\frac{1}{2}$	30. $\frac{1}{14}$
	16. $\frac{5}{12}$	31. $\frac{7}{24}$
	17. $\frac{7}{10}$	32. $\frac{13}{30}$
	18. $\frac{7}{12}$	33. $\frac{33}{100}$
	19. $\frac{1}{6}$	34. 1
	20. $\frac{11}{12}$	35. 4
	21. $\frac{7}{18}$	36. 3
	22. $\frac{3}{20}$	37. 0
		38. 9
		39. 6
		40. 7

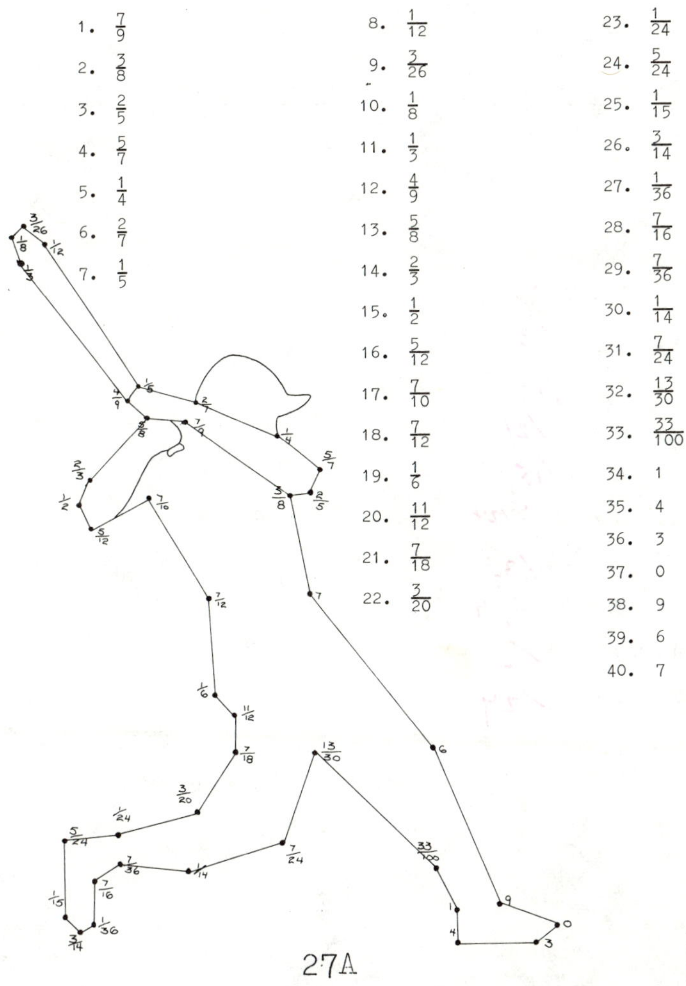

27A

A winner of two Nobel Prizes:

$$\underset{19\ 6\ 11\ 4\ 13}{M\ A\ R\ I\ E} \quad \underset{1\ 7\ 10\ 24\ 17\ 24\ 18\ 1\ 7\ 6}{S\ K\ L\ O\ D\ O\ W\ S\ K\ A} \quad \underset{5\ 3\ 11\ 4\ 13}{C\ U\ R\ I\ E}$$

1. L = 10   3/10
2. W = 18   1/18
3. A = 6   1/6
4. D = 17   1/24
5. E = 13   1/30
6. K = 7   1/36
7. R = 11   1/60
8. U = 3
9. S = 1   1/24
10. I = 4   1/25
11. O = 24   15/24
12. M = 19   1/24
13. C = 5   5/6

28A

Founder of Modern Astronomy:

$$\frac{N}{9}\ \frac{I}{5}\ \frac{C}{22}\ \frac{HO}{4}\ \frac{L}{24}\ \frac{A}{3}\ \frac{S}{8}\ \frac{}{1}\qquad \frac{C}{22}\ \frac{O}{24}\ \frac{P}{7}\ \frac{E}{10}\ \frac{R}{2}\ \frac{N}{9}\ \frac{I}{5}\ \frac{C}{22}\ \frac{U}{14}\ \frac{S}{1}$$

1. L = 3
2. R = 2
3. H = 4/15
4. S = 1/10
5. I = 5/18
6. E = 10/21
7. A = 8/15
8. O = 24 5/24
9. U = 14 /33
10. N = 9 5/9
11. C = 22 /27
12. P = 7 /24

29A

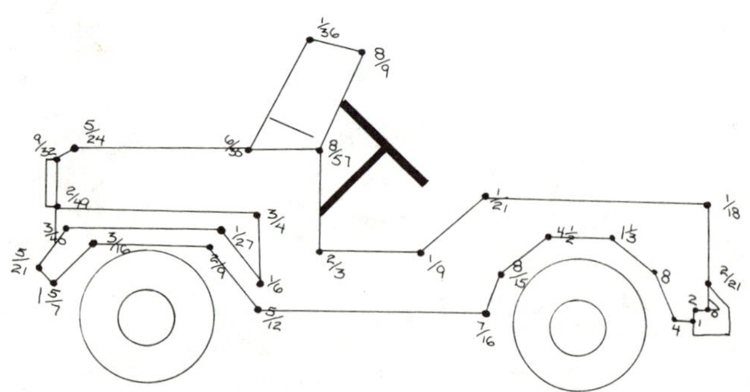

1. 4	11. $\frac{5}{21}$	21. $\frac{8}{9}$
2. 8	12. $\frac{3}{40}$	22. $\frac{8}{57}$
3. $1\frac{1}{3}$	13. $\frac{1}{27}$	23. $\frac{2}{3}$
4. $4\frac{1}{2}$	14. $\frac{1}{6}$	24. $\frac{1}{9}$
5. $\frac{8}{15}$	15. $\frac{3}{4}$	25. $\frac{1}{21}$
6. $\frac{7}{16}$	16. $\frac{2}{49}$	26. $\frac{1}{18}$
7. $\frac{5}{12}$	17. $\frac{9}{32}$	27. $\frac{2}{21}$
8. $\frac{2}{9}$	18. $\frac{5}{24}$	28. 0
9. $\frac{3}{16}$	19. $\frac{6}{35}$	29. 2
10. $1\frac{5}{7}$	20. $\frac{1}{36}$	30. 1

30A

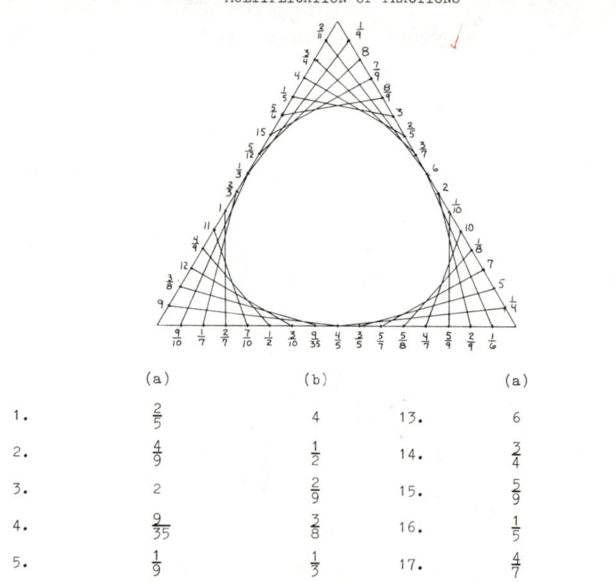

	(a)	(b)		(a)	(b)
1.	$\frac{2}{5}$	4	13.	6	$\frac{1}{6}$
2.	$\frac{4}{9}$	$\frac{1}{2}$	14.	$\frac{3}{4}$	$\frac{3}{7}$
3.	2	$\frac{2}{9}$	15.	$\frac{5}{9}$	$\frac{1}{10}$
4.	$\frac{9}{35}$	$\frac{3}{8}$	16.	$\frac{1}{5}$	3
5.	$\frac{1}{9}$	$\frac{1}{3}$	17.	$\frac{4}{7}$	10
6.	5	$\frac{3}{5}$	18.	$\frac{5}{12}$	8
7.	$\frac{1}{7}$	$\frac{2}{3}$	19.	$\frac{3}{10}$	12
8.	$\frac{7}{9}$	15	20.	7	$\frac{5}{7}$
9.	$\frac{1}{8}$	5	21.	$\frac{7}{10}$	11
10.	$\frac{8}{9}$	$\frac{5}{6}$	22.	$\frac{4}{5}$	9
11.	1	$\frac{2}{7}$	23.	$\frac{1}{3}$	$\frac{9}{10}$
12.	$\frac{4}{5}$	$\frac{1}{4}$	24.	$\frac{2}{11}$	6

31A

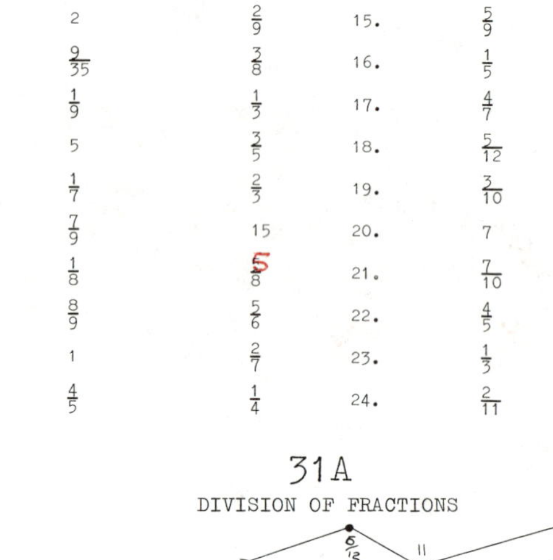

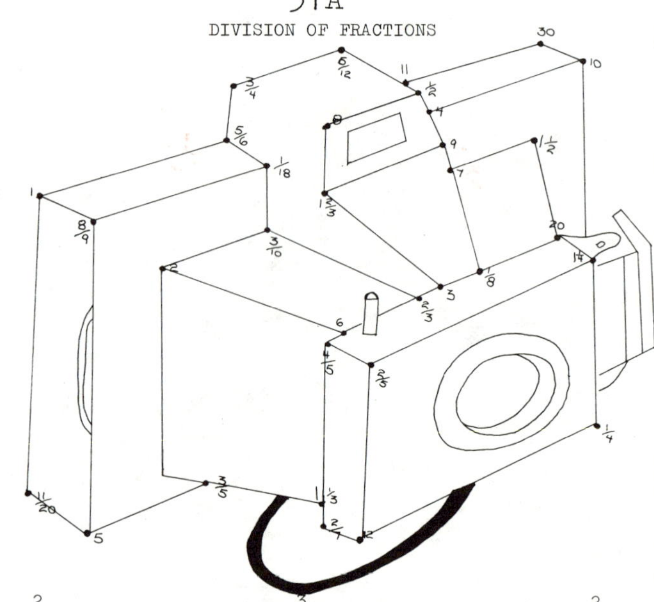

1. $\frac{2}{3}$		12. $\frac{3}{5}$		23. $1\frac{2}{3}$
2. $\frac{3}{10}$		13. 5		24. 3
3. 2		14. $\frac{11}{20}$		25. $\frac{1}{8}$
4. 6		15. 1		26. 20
5. $\frac{4}{5}$		16. $\frac{8}{9}$		27. $1\frac{1}{2}$
6. $\frac{2}{5}$		17. $\frac{1}{18}$		28. 7
7. $1\frac{1}{4}$		18. $\frac{5}{6}$		29. 9
8. $\frac{1}{4}$		19. $\frac{3}{4}$		30. 4
9. 12		20. $\frac{5}{12}$		31. 10
10. $\frac{2}{7}$		21. $\frac{1}{2}$		32. 30
11. $1\frac{1}{3}$		22. 8		33. 11

32A

DIVISION OF FRACTIONS

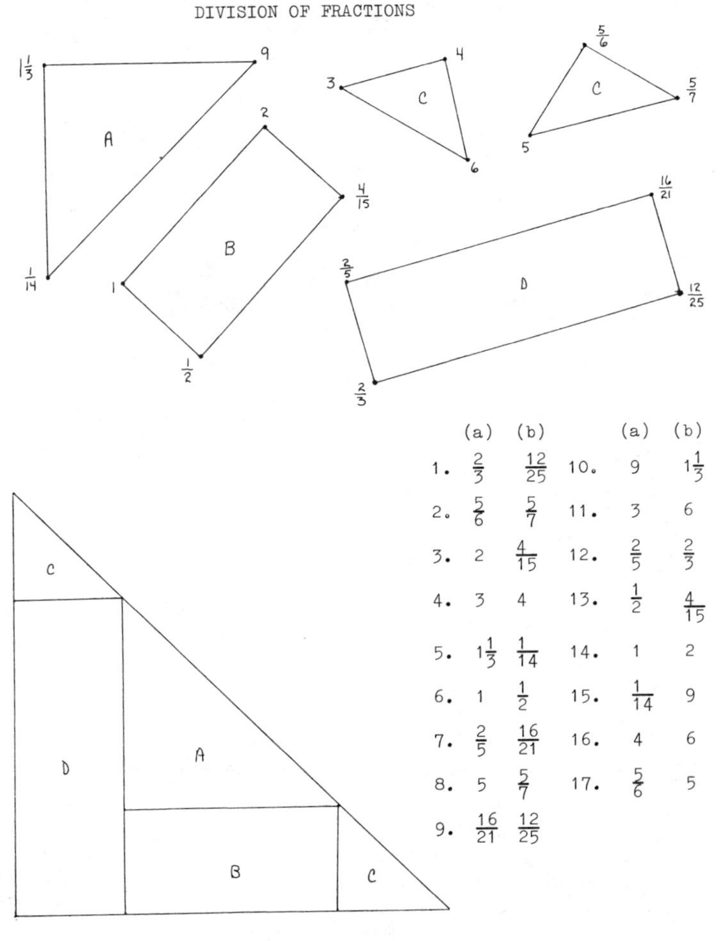

33A

	(a)	(b)		(a)	(b)
1.	$\frac{2}{3}$	$\frac{12}{25}$	10.	9	$1\frac{1}{3}$
2.	$\frac{5}{6}$	$\frac{5}{7}$	11.	3	6
3.	2	$\frac{4}{15}$	12.	$\frac{2}{5}$	$\frac{2}{3}$
4.	3	4	13.	$\frac{1}{2}$	$\frac{4}{15}$
5.	$1\frac{1}{3}$	$\frac{1}{14}$	14.	1	2
6.	1	$\frac{1}{2}$	15.	$\frac{1}{14}$	9
7.	$\frac{2}{5}$	$\frac{16}{21}$	16.	4	6
8.	5	$\frac{5}{7}$	17.	$\frac{5}{6}$	5
9.	$\frac{16}{21}$	$\frac{12}{25}$			

FRACTION OPERATIONS

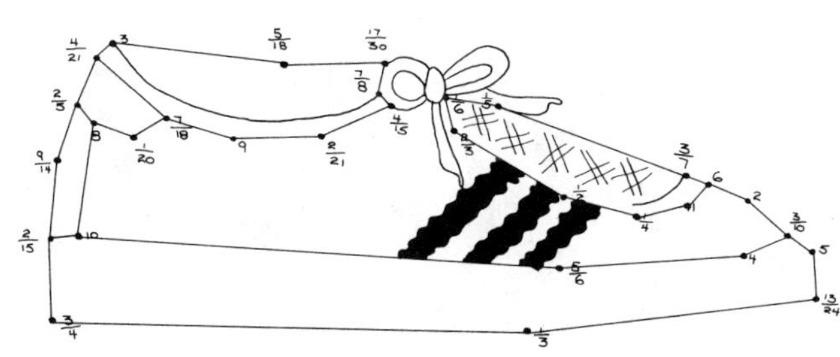

34A

1.	$\frac{3}{7}$	12.	$\frac{5}{6}$	23.	3
2.	$\frac{1}{5}$	13.	10	24.	$\frac{4}{21}$
3.	$\frac{1}{6}$	14.	8	25.	$\frac{2}{5}$
4.	$\frac{2}{3}$	15.	$\frac{1}{20}$	26.	$\frac{9}{14}$
5.	$\frac{1}{2}$	16.	$\frac{7}{18}$	27.	$\frac{2}{15}$
6.	$\frac{1}{4}$	17.	9	28.	$\frac{3}{4}$
7.	1	18.	$\frac{2}{21}$	29.	$\frac{1}{3}$
8.	6	19.	$\frac{4}{15}$	30.	$\frac{13}{24}$
9.	2	20.	$\frac{7}{8}$	31.	5
10.	$\frac{3}{10}$	21.	$\frac{17}{30}$	32.	$\frac{3}{10}$
11.	4	22.	$\frac{5}{18}$		

FRACTION OPERATIONS

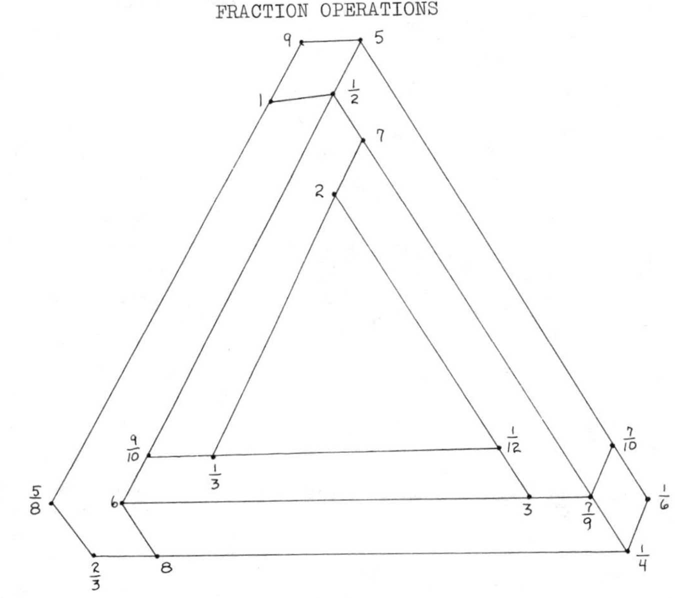

35A

	(a)	(b)		(a)	(b)
1.	$\frac{1}{2}$	1	9.	5	9
2.	$\frac{5}{8}$	$\frac{2}{3}$	10.	$\frac{1}{6}$	5
3.	$\frac{7}{9}$	$\frac{7}{10}$	11.	$\frac{9}{10}$	$\frac{1}{12}$
4.	2	3	12.	$\frac{1}{3}$	7
5.	$\frac{1}{6}$	$\frac{1}{4}$	13.	6	$\frac{7}{9}$
6.	6	5	14.	$\frac{2}{3}$	$\frac{1}{4}$
7.	6	8	15.	9	$\frac{5}{8}$
8.	$\frac{1}{2}$	$\frac{1}{4}$			

READING AND WRITING DECIMALS

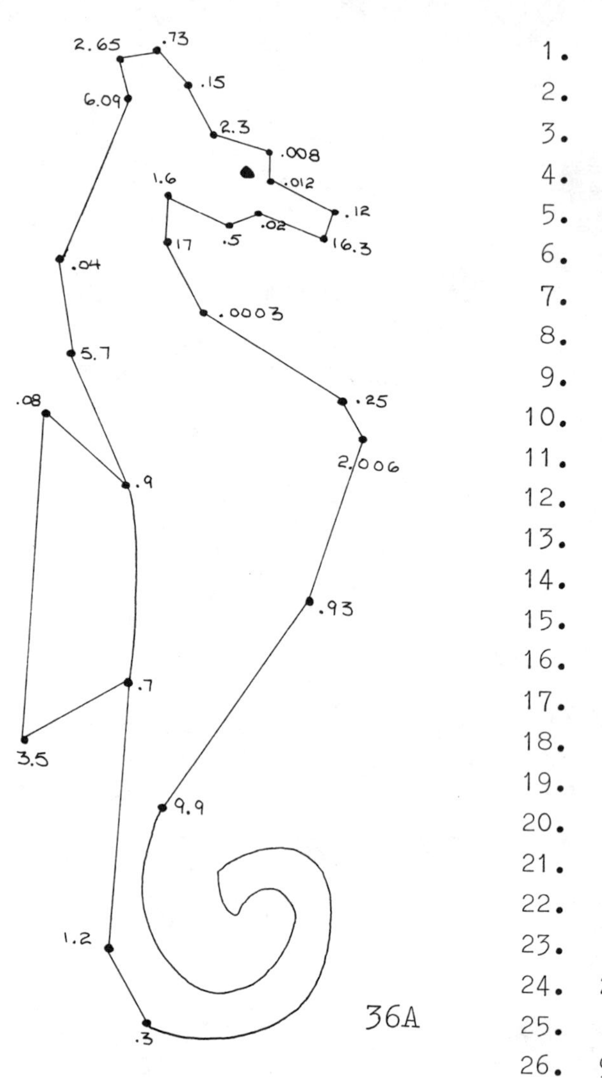

1. .3
2. 1.2
3. .7
4. 3.5
5. .08
6. .9
7. 5.7
8. .04
9. 6.09
10. 2.65
11. .73
12. .15
13. 2.3
14. .008
15. .012
16. .12
17. 16.3
18. .02
19. .5
20. 1.6
21. 17
22. .0003
23. .25
24. 2.006
25. .93
26. 9.9

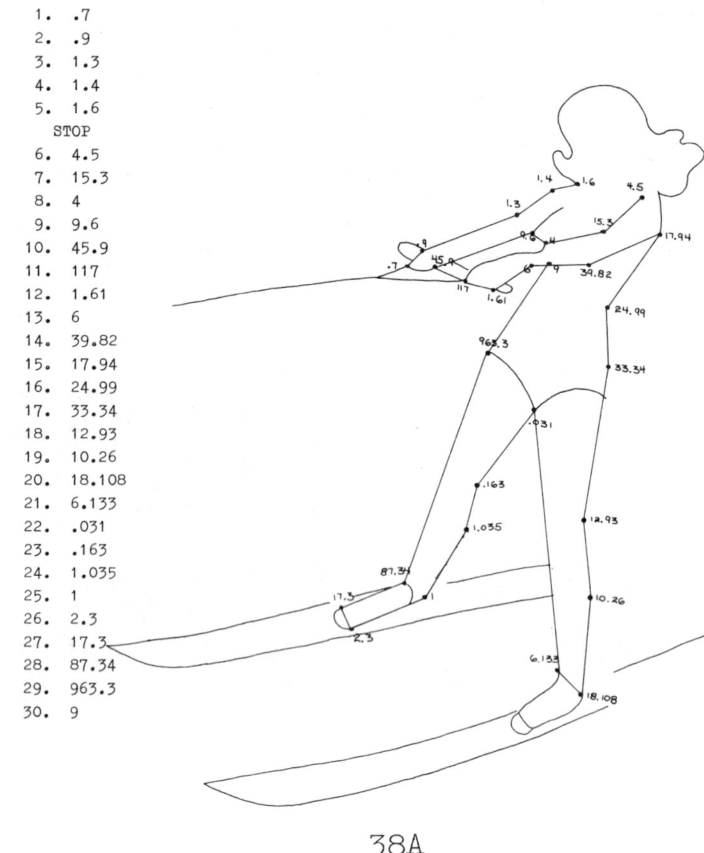

1. .7
2. .9
3. 1.3
4. 1.4
5. 1.6
STOP
6. 4.5
7. 15.3
8. 4
9. 9.6
10. 45.9
11. 117
12. 1.61
13. 6
14. 39.82
15. 17.94
16. 24.99
17. 33.34
18. 12.93
19. 10.26
20. 18.108
21. 6.133
22. .031
23. .163
24. 1.035
25. 1
26. 2.3
27. 17.3
28. 87.34
29. 963.3
30. 9

FRACTIONS AND DECIMALS

.543 $\frac{1}{9}$ $\frac{1}{4}$ $\frac{11}{100}$	.715 .25 $\frac{9}{100}$ $\frac{2}{5}$	$\frac{5}{6}$ .09 $\frac{7}{50}$ .17	.92 .14 .5 $\frac{3}{50}$	$\frac{1}{12}$ $\frac{1}{2}$ .52 $\frac{53}{100}$
.11 $\frac{1}{3}$ .1 $\frac{97}{100}$	.4 $\frac{1}{10}$ $\frac{3}{1000}$ .02	$\frac{17}{100}$ .003 .01 $\frac{3}{4}$	.06 $\frac{1}{100}$ .9 .0017	.53 $\frac{9}{10}$ $\frac{1}{6}$ .061
.97 .432 .125 $\frac{73}{100}$	$\frac{1}{50}$ $\frac{1}{8}$ .89 .017	.75 $\frac{89}{100}$ .07 $\frac{3}{10}$	$\frac{17}{10000}$ $\frac{7}{100}$ .6 $\frac{3}{8}$	$\frac{61}{1000}$ $\frac{3}{5}$ .34 $\frac{63}{100}$
.73 .655 .131 $\frac{7}{8}$	$\frac{17}{1000}$ $\frac{131}{1000}$ .2 .7	.3 $\frac{1}{5}$ $\frac{423}{1000}$ .07	.375 .423 $\frac{3}{10000}$ $\frac{4}{5}$	.63 .0003 $\frac{2}{9}$ .625
.875 $\frac{8}{11}$ .05 $\frac{2}{3}$	$\frac{7}{10}$ $\frac{1}{20}$ .001 $\frac{4}{9}$	$\frac{7}{100}$ $\frac{1}{1000}$ .04 $\frac{2}{7}$	.8 $\frac{1}{25}$ .29 .123	$\frac{5}{8}$ $\frac{29}{100}$ $\frac{5}{7}$ .617

37A

ADDITION OF DECIMALS

Prime Minister of India:

$$\frac{I}{7} \frac{N}{.5} \frac{D}{8} \frac{I}{7} \frac{R}{3.0} \frac{A}{4} \qquad \frac{G}{.3} \frac{A}{4} \frac{N}{.5} \frac{D}{8} \frac{H}{6} \frac{I}{7}$$

Former Prime Minister of Israel:

$$\frac{G}{.3} \frac{O}{.4} \frac{L}{.1} \frac{D}{8} \frac{A}{4} \qquad \frac{M}{4.9} \frac{E}{.2} \frac{I}{7} \frac{R}{3.0}$$

1. $\frac{6}{H}.\underline{1}$

2. $\underline{8}\ \underline{0}.\underline{\underline{2}}{E}$

3. $\underline{2}\ \frac{4}{A}.\underline{1}$

4. $.\underline{\underline{5}}{N}\ \underline{2}$

5. $\underline{1}\ \underline{0}.\underline{\underline{1}}{L}$

6. $\underline{3}\ \underline{3}.\underline{0}{R}\ \underline{2}$

7. $\underline{4}\ \underline{2}\ \underline{1}.\frac{4}{0}\ \underline{2}$

8. $\frac{7}{I}.\underline{1}$

9. $\frac{8}{D}.\underline{5}\ \underline{1}\ \underline{4}$

10. $\underline{1}\ \underline{2}.\frac{3}{G}$

11. $\underline{2}\ \underline{4}.\underline{9}{M}\ \underline{4}$

39A

SUBTRACTION OF DECIMALS

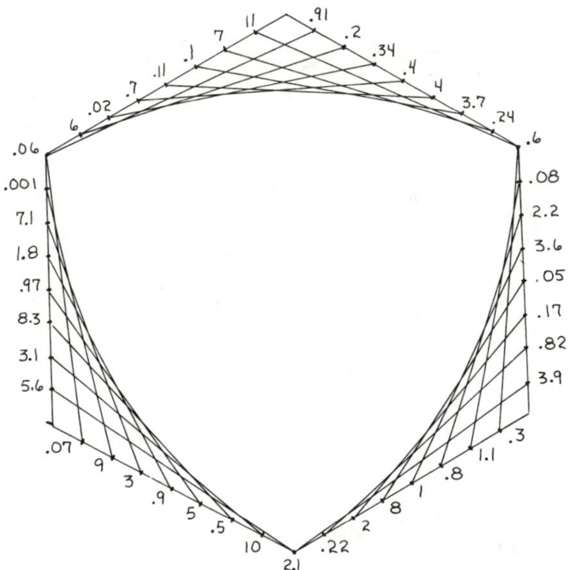

	(a)	(b)
1.	.7	.4
2.	.6	.3
3.	.9	1.8
4.	2.1	3.9
5.	.24	7
6.	1	3.6
7.	.2	6
8.	3.1	10
9.	.06	.07
10.	.17	2
11.	7.1	3
12.	.91	.06
13.	1.1	.08
14.	4	.11
15.	.5	8.3
16.	2.2	.8
17.	.97	5
18.	3.7	.1
19.	.001	9
20.	.82	.22
21.	2.1	5.6
22.	8	.05
23.	.02	.34
24.	11	.6

40A

SUBTRACTION OF DECIMALS

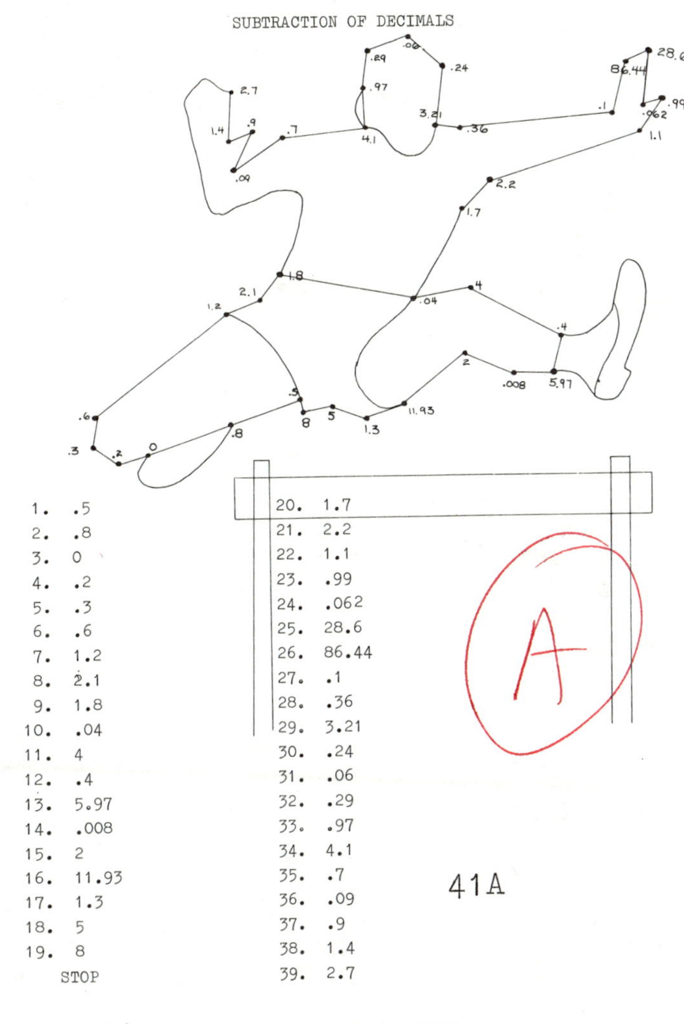

1.	.5	20.	1.7
2.	.8	21.	2.2
3.	0	22.	1.1
4.	.2	23.	.99
5.	.3	24.	.062
6.	.6	25.	28.6
7.	1.2	26.	86.44
8.	2.1	27.	.1
9.	1.8	28.	.36
10.	.04	29.	3.21
11.	4	30.	.24
12.	.4	31.	.06
13.	5.97	32.	.29
14.	.008	33.	.97
15.	2	34.	4.1
16.	11.93	35.	.7
17.	1.3	36.	.09
18.	5	37.	.9
19.	8	38.	1.4
STOP		39.	2.7

41A

MULTIPLICATION OF DECIMALS

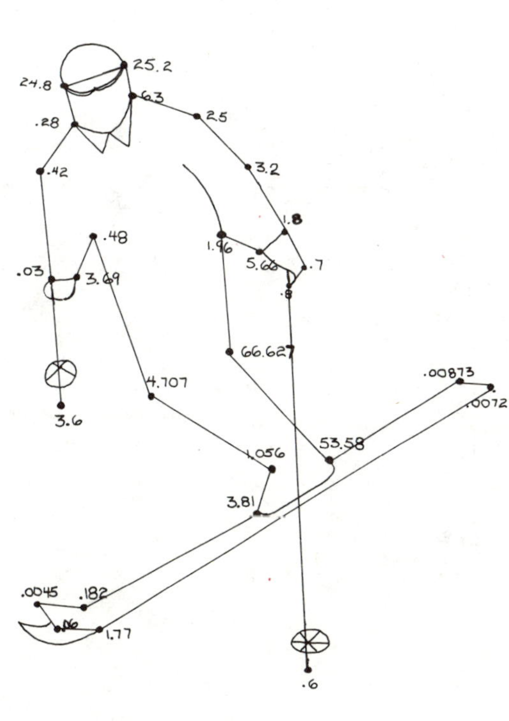

1.	.6
2.	.8
3.	.7
4.	1.8
5.	3.2
6.	2.5
7.	6.3
8.	25.2
9.	24.8
10.	.28
11.	.42
12.	.03
13.	3.6
STOP	
14.	3.69
15.	.48
16.	4.707
17.	1.056
18.	3.81
19.	.182
20.	.0045
21.	.06
22.	1.77
23.	.0072
24.	.00873
25.	53.58
26.	66.627
27.	1.96
28.	5.66

42A

MULTIPLICATION OF DECIMALS

D	E	C	I	M	A	L	S		A	R	E		P	O	I	N	T	L	E	S	S	?
1.0	11	2	.04	5	.8	4	7		.8	.01	11		3.1	.3	.04	.12	.13	4	11	7	7	

1. .3 / O
2. 2.8 / A
3. .1 2 / N
4. 1 1.2 / E
5. .1 3 1 5 / T
6. .0 1 8 9 / R
7. 7.5 6 / S
8. 2.3 6 8 / C
9. 5.3 1 3 1 / M
10. 4.7 3 / L
11. .0 4 6 7 2 / I
12. 1.0 8 1 5 / D
13. 5 3.1 9 6 / P

43A

DIVISION OF DECIMALS

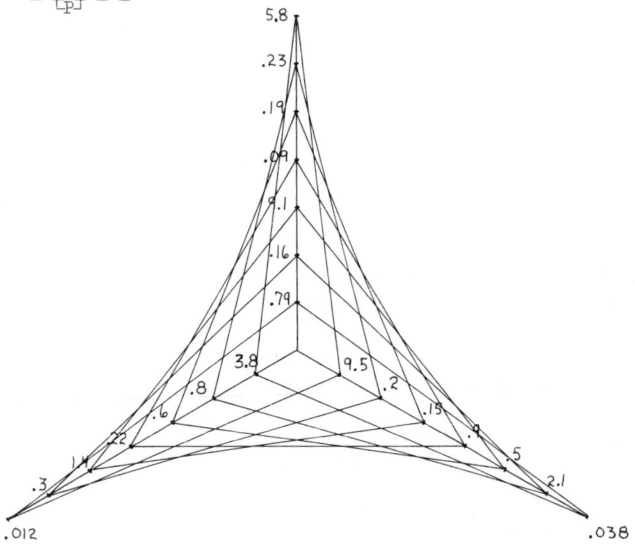

5.8
.23
.19
.09
9.1
.16
.79
3.8
9.5
.8
.2
.6
.15
.22
.9
1.4
.5
.3
2.1
.012
.038

	(a)	(b)
1.	.2	.3
2.	.6	.5
3.	.9	.09
4.	.8	.23
5.	.15	1.4
6.	2.1	.16
7.	1.4	9.1
8.	9.5	.012
9.	.038	3.8
10.	5.8	9.5
11.	.79	.038
12.	.6	.19
13.	.3	.16
14.	.22	.9
15.	.8	2.1
16.	9.1	.5
17.	5.8	3.8
18.	.012	.79
19.	.09	.22
20.	.23	.2
21.	.19	.15

44A

DIVISION OF DECIMALS

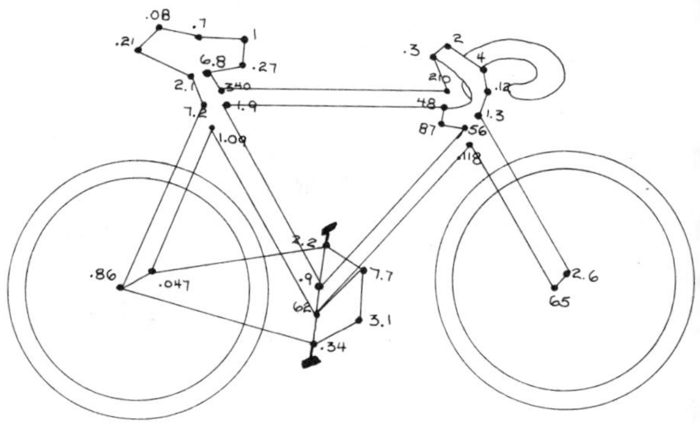

.08
.7
.21
1
6.8
.27
.3
2
4
2.1
340
210
.12
7.2
1.9
48
1.3
1.09
87
56
.118
2.2
.86
.047
.9
7.7
2.6
65
62
3.1
.34

1. .7
2. .08
3. .21
4. 2.1
5. 7.2
6. .86
7. .34
8. 3.1
9. 7.7
10. 62
11. 1.09
12. .047
13. 2.2
14. .9
15. 1.9
16. 48
17. 87
18. 56
19. .9
20. 62
21. .118
22. 65
23. 2.6
24. 1.3
25. .12
26. 4
27. 2
28. .3
29. 210
30. 340
31. 6.8
32. .27
33. 1
34. .7

45A

DECIMAL OPERATIONS

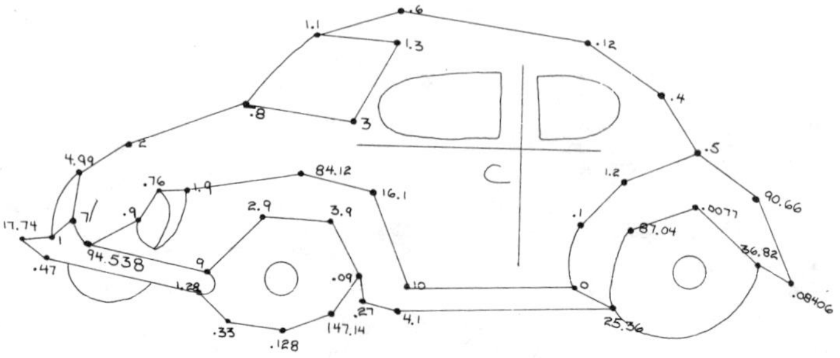

1.1
.6
1.3
.12
.4
.8
3
4.99
2
.5
.76
1.9
84.12
16.1
1.2
C
2.9
3.9
.0077
90.66
17.74
7
.9
.1
87.04
94.538
9
36.82
.47
1.28
.09
10
0
.08406
.27
4.1
147.14
.33
.128
25.36

1. .5
2. .4
3. .12
4. .6
5. 1.1
6. 1.3
7. 3
8. .8
9. 2
10. 4.99
11. 7
12. 1
13. 17.74
14. .47
15. 1.28
16. .33
17. .128
18. 147.14
19. .09
20. .27
21. 4.1
22. 25.36
23. 0
24. 10
25. 16.1
26. 84.12
27. 1.9
28. .76
29. .9
30. 94.538
31. 9
32. 2.9
33. 3.9
34. .09

STOP

35. 87.04
36. .0077
37. 36.82
38. .08406
39. 90.66
40. .5
41. 1.2
42. .1

46A

DECIMAL OPERATIONS

47A

1	1	6	.3	■	5	7	.9	5
2	7	.2	■	.2	■	.1	1	8
.5	7	■	.7	9	3	■	1	.1
3	■	.1	4	■	5	.7	■	6
■	.3	5	■	.8	■	1	2	■
8	■	7	.4	■	.8	4	■	9
.8	4	■	7	.7	4	■	3	.2
7	.3	8	■	9	■	5	.3	3
9	6	.9	8	■	1	.3	1	1

ROMAN NUMERALS

48A

X	I	I	■	C	■	L	I	V
I	X	■	C	L	I	■	V	I
X	■	D	C	X	I	X	■	I
■	C	C	■	V	■	X	X	■
X	X	X	V	■	M	X	V	I
■	X	X	■	M	■	V	I	■
M	■	X	X	X	V	I	■	D
D	C	■	C	C	I	■	M	C
C	I	X	■	V	■	D	X	C

49A

52 LVII 74 CDXII	47 LXXIV 17 CIV	IV XVII 105 MXC	MLV CV DCIX CCCX	M 609 87 MMDIX
412 DIII 1600 MMLIV	104 MDC MI VI	1090 1001 60 DCXII	310 LX 1400 XXIV	2509 MCD 103 CDX
2054 33 DLIV 49	·6 554 LIX XXVI	612 59 110 MCX	24 CX DIX 69	410 509 CIII MCCX
XLIX L 90 CMXC	26 XC 40 19	1110 XL CC CCXC	LXIX 200 DVI CLX	1210 506 93 21
990 73 LVI 3	XIX 56 CVI 76	290 106 526 LXXX	160 DXXVI MM 83	XXI 2000 8 MCCC

BASIC: PRINT, GOTO

WHEN STUDYING

TRY

THINKING

50A

BASIC: PRINT, LET, STRINGS

KNOCK KNOCK! WHO' THERE?

SARAH WHO

SARAH DOCTOR IN THE HOUSE

KENNETH WHO

KENNETH B RIGHT

CANDY WHO

CANDY TEST BE POSTPONED

ELSIE WHO

ELSIE U LATER

LIZETTE WHO

LIZETTE ALL THERE IS

51A

BASIC: PRINT, LET, STRINGS

How old is Bob?   12

       Ellen?   15

       Harry?   15

       Mary?   10

        Nick?   24

52A

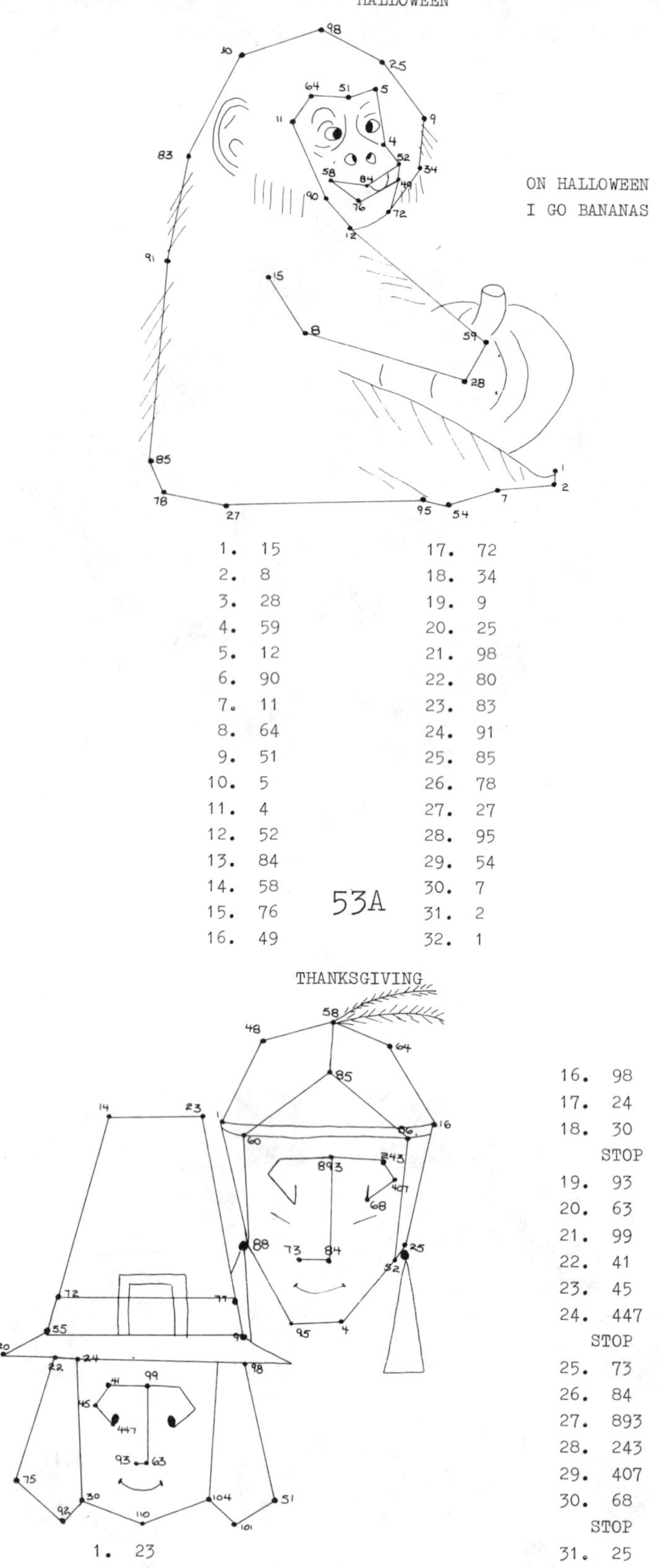

1.	15	17.	72
2.	8	18.	34
3.	28	19.	9
4.	59	20.	25
5.	12	21.	98
6.	90	22.	80
7.	11	23.	83
8.	64	24.	91
9.	51	25.	85
10.	5	26.	78
11.	4	27.	27
12.	52	28.	95
13.	84	29.	54
14.	58	30.	7
15.	76	31.	2
16.	49	32.	1

53A

1.	23	16.	98
2.	14	17.	24
3.	72	18.	30
			STOP
4.	77	19.	93
5.	9	20.	63
6.	55	21.	99
7.	20	22.	41
8.	22	23.	45
9.	75	24.	447
			STOP
10.	92	25.	73
11.	30	26.	84
12.	110	27.	893
13.	104	28.	243
14.	101	29.	407
15.	51	30.	68
			STOP
		31.	25
		32.	16
		33.	64
		34.	58
		35.	48
		36.	1
		37.	88
		38.	95
		39.	4
		40.	52
		41.	56
		42.	85
		43.	60
		44.	88

54A

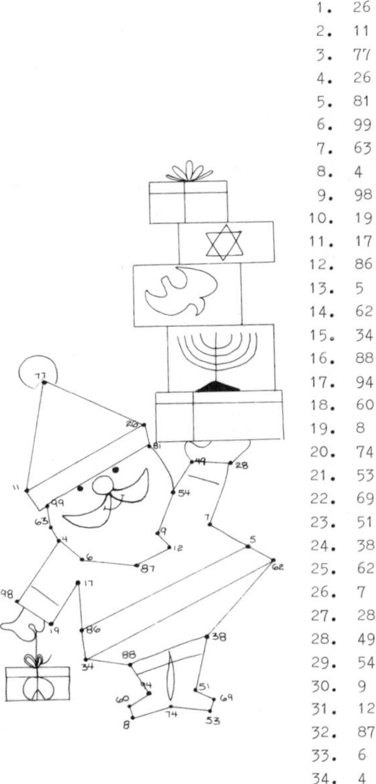

1. 26
2. 11
3. 77
4. 26
5. 81
6. 99
7. 63
8. 4
9. 98
10. 19
11. 17
12. 86
13. 5
14. 62
15. 34
16. 88
17. 94
18. 60
19. 8
20. 74
21. 53
22. 69
23. 51
24. 38
25. 62
26. 7
27. 28
28. 49
29. 54
30. 9
31. 12
32. 87
33. 6
34. 4

55A

VALENTINES DAY

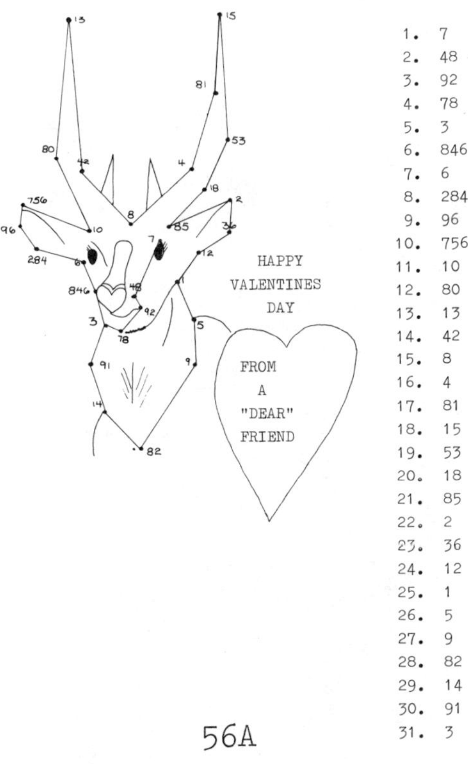

1. 7
2. 48
3. 92
4. 78
5. 3
6. 846
7. 6
8. 284
9. 96
10. 756
11. 10
12. 80
13. 13
14. 42
15. 8
16. 4
17. 81
18. 15
19. 53
20. 18
21. 85
22. 2
23. 36
24. 12
25. 1
26. 5
27. 9
28. 82
29. 14
30. 91
31. 3

56A

EASTER

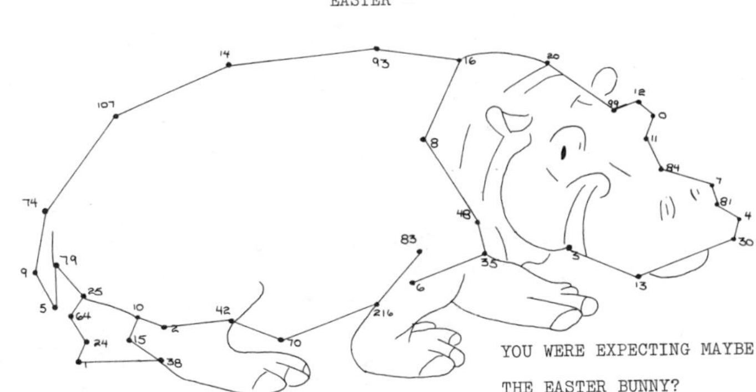

1. 6	20. 2
2. 35	21. 42
3. 48	22. 70
4. 8	23. 216
5. 16	24. 83
6. 93	STOP
7. 14	25. 3
8. 107	26. 13
9. 74	27. 30
10. 9	28. 4
11. 5	29. 81
12. 79	30. 7
13. 25	31. 84
14. 64	32. 11
15. 24	33. 0
16. 1	34. 12
17. 38	35. 99
18. 15	36. 20
19. 10	

57A